T0372027

Hans Jansson, PhD

Industrial Products: A Guide to the International Marketing Economics Model

Pre-publication REVIEWS, COMMENTARIES, EVALUATIONS . . .

"This book is A WELCOME ADDITION TO THE LITERATURE ON BUSINESS-TO-BUSINESS AND INTERNATIONAL MARKETING, particularly with its emphasis on linkage processes and marketing strategies within the ASEAN countries. I found the case study analyses valuable, as a way of clearly identifying the processes at work as the European TNCs (the focus of the study) sought to penetrate markets with considerable cross-cultural diversity. Researchers will also appreciate the extended work on development of a theoretical framework for such business-to-business marketing

pre-publication
REVIEWS, COMMENTARIES, EVALUATIONS . . .

activities. There are numerous practical examples of the marketing approaches to the different markets in Southeast Asia used by the TNCs which provide a rich source for those teaching in the field–with a strong emphasis on the longer-term commitment required for success. All in all, A VALUABLE SOURCE BOOK FOR LECTURERS AND RESEARCHERS IN THE FIELD."

Lawrence S. Welch, PhD
Graduate School of Management,
Monash University,
Clayton, Australia

International Business Press
An Imprint of The Haworth Press, Inc.

Industrial Products

A Guide to the International Marketing Economics Model

INTERNATIONAL BUSINESS PRESS
Erdener Kaynak, PhD
Executive Editor

New, Recent, and Forthcoming Titles:

International Business Handbook edited by V. H. (Manek) Kirpalani

Sociopolitical Aspects of International Marketing edited by Erdener Kaynak

How to Manage for International Competitiveness edited by Abbas J. Ali

International Business Expansion into Less-Developed Countries: The International Finance Corporation and Its Operations by James C. Baker

Product-Country Images: Impact and Role in International Marketing edited by Nicolas Papadopoulos and Louise A. Heslop

The Global Business: Four Key Marketing Strategies edited by Erdener Kaynak

Multinational Strategic Alliances edited by Refik Culpan

Market Evolution in Developing Countries: The Unfolding of the Indian Market by Subhash C. Jain

A Guide to Successful Business Relations with the Chinese: Opening the Great Wall's Gate by Huang Quanyu, Richard Andrulis and Chen Tong

Industrial Products: A Guide to the International Marketing Economics Model by Hans Jansson

Euromarketing: Effective Strategies for International Trade and Export edited by Erdener Kaynak and Pervez N. Ghauri

Industrial Products

A Guide to the International Marketing Economics Model

Hans Jansson, PhD

International Business Press
An Imprint of The Haworth Press, Inc.
New York • London • Norwood (Australia)

Published by

International Business Press, an imprint of The Haworth Press, Inc., 10 Alice Street, Binghamton, NY 13904-1580

Library of Congress Cataloging-in-Publication Data

Jansson, Hans.
Industrial products : a guide to the international marketing economics model / Hans Jansson.
p. cm.
Includes bibliographical references and index.
ISBN 1-56024-425-9 (alk. paper).
1. Export marketing–Management. 2. Industrial marketing–Management. 3. International business enterprises–Management. I. Title.
HF1416.J36 1993
658.8′48–dc 20 93-15578
CIP

CONTENTS

ABOUT THE AUTHOR

Hans Jansson, PhD, is Professor of marketing in the School of Economics and Business Administration at Lund University. He has extensive research and teaching experience in the marketing and international business fields, mainly from Swedish universities, but also from a one-year senior visiting fellowship at the National University of Singapore. Dr. Jansson has stayed longer periods in Asia during the past fifteen years doing field research on European transnational corporations. He is the author of several books and articles on a variety of international business topics related to business in Asia.

List of Tables and Figures

Tables

Figures

Foreword

We are all talking about the urgency of interdisciplinary approaches to the professional areas of business administration. Admonitions to do more research in the interface of disciplines are especially common in marketing; that is natural in view of its ancestry in economics and behavioral sciences, and marketing's position as the prime boundary-spanning function of the firm. Too often, however, we hear too much talk and see too little action.

Hans Jansson is to be commended for taking the challenge seriously. His work on international industrial marketing consummates the marriage of transactions cost economics and organization theory, with an occasional outreach into political science. His first round of chapters lays a solid foundation of theory. The perspective thus developed is applied in the next round, in which the experience of a dozen Swedish multinationals in Southeast Asian countries is examined in detail. The final chapter is devoted to a discussion of managerial implications of this outstanding research project.

True, no numbers are presented, thus inviting skepticism among those American scholars to whom bean-counting is the sacred way of doing research. But to those of us aware of the lacunas in basic theory in the social sciences, the frontal approach taken by Jansson in theory development is not only ambitious, but, to me, also quite promising. Taking into account the problems of field research in international business–and particularly in the LDC and NIC–in-depth qualitative exploration as evidenced here is simply indispensable preparation on the way to research based on statistical analysis, quite apart from the substantial intrinsic value of empirical exploration as presented here.

At the theory level, Jansson is justly fascinated by the *vertical* aspects of industrial marketing, bringing to the fore the challenge to both buyer and seller of defining (dynamically) their positions in the value-added chain. This raises the specter of the make-or-buy decision (modestly reflected at the consumer level by do-it-yourself or buy-it). He points out that these aspects have been pretty well neglected in the industrial marketing literature to date–with the notable exception of distribution channels. The importance of the single transaction in this area stimulates linkages and networks–between markets and hierarchies–everywhere. A key difference between such cooperative arrangements in industrialized and developing countries is that whereas networking in the former is essentially inter-organizational, it is essentially interpersonal in the latter.

Jansson's industrial marketing paradigm, which may be viewed appropriately as an inter-organizational system theory, has the great merit of potential universal application, due to its broad scope and inherent flexibility. He is, however, careful to emphasize that the empirical part of the study lends immediate support to the validity of his theoretical propositions limited to the Southeast Asian context. Hans Jansson is keenly aware that the ultimate test of relevance of research in professional disciplines–be they related to business, medicine, or technology–is its usefulness to practitioners. His last chapter on managerial implications of his work to multinationals actually or potentially active in the Southeast Asian region sets a commendable example. Given my own experience in Thailand, the PRC, Africa, and Latin America, I believe that several of his prescriptions would (nay, should!) be of equal interest in many less developed and industrializing countries outside that region.

Hans B. Thorelli
Distinguished Professor of Business Administration Em.
Indiana University

Preface

The international marketplace is fascinating to students. But they need tools to get an understanding of the variety and variability of international markets. There is a standard international (or multinational, or global) marketing textbook structure telling us how to formulate and implement international (multinational, global) marketing plans. We are told that international marketing planning should consider the economic, cultural, political, and legal dimensions of the international (multinational, global) marketing environment of the firm. These books, usually comprising 600-800 pages, provide students with a great number of concepts that, unfortunately, filter out much of the variety and variability of the international marketplace.

The alternative to textbooks is real-life stories by experienced managers. Such stories capture much of the exciting features of international marketing, but the understanding they convey is related to the specific situation and can seldom be generalized. They usually fall back on ad-hoc explanations based on cultural or institutional peculiarities of the country or the market.

Hans Jansson's study of "International Marketing of Industrial Products in Southeast Asia" is an attempt to do something in between, to catch something of the cultural and institutional complexities of the Southeast Asian region without sacrificing general understanding of marketing strategy. He develops a theoretical framework–the marketing economics approach–allowing him to view culture and institutions as integral parts of the markets. Based on this approach he presents rigorous empirical research of transnational firms' industrial marketing strategies in the Southeast Asian countries. The reader meets an ambitious work providing significant insights into the interplay between marketing strategies and institutions in an interesting region.

Personally, I agree strongly when he places attention on the linkage strategies describing how the firms develop and maintain links to the customers. And I find this discussion of linkage strategies in the context of the cultural segmentation in the Southeast Asian markets very interesting. In particular, he provides empirical data about the establishment and use of contact nets really giving a scent of the international marketplace. This is a promising avenue for international marketing studies.

Jan Johanson
Professor of International Business
Department of Business Studies
Uppsala University

Acknowledgements

This book is due in large part to the kind cooperation of the European transnational corporations studied. Persons interviewed at these companies both at the headquarters in Europe and at the subsidiaries in Southeast Asia have shared their knowledge and experience with me during numerous visits between 1984 and 1991. I want to express my deep gratitude to these people and organizations but for whose direct and indirect help the completion of the study reported in this book would not have been possible. I also want to thank many of my colleagues who have commented on the different versions of the book manuscript. I am especially grateful to Associate Professor Ingmar Tufvesson at Lund University, and Dr. Philip L. Dawes and Associate Professor Pang Eng Fong at the National University of Singapore for their valuable suggestions regarding the contents of the book. Special thanks and appreciation go to Associate Professor Robert Goldsmith at Lund University for his highly professional correction of my English and many valuable comments on the manuscript.

Of course, my family deserves a special mention. I extend my warmest thanks to my wife Carina, and to Erik and Emma. As in the past, they were again very understanding of the demands of a scholarly pursuit of this nature and were a constant source of encouragement and support both at home and during our one-year stay in Singapore.

NOTES FOR PROFESSIONAL LIBRARIANS AND LIBRARY USERS

This is an original book title published by International Business Press, an imprint of The Haworth Press, Inc. Unless otherwise noted in specific chapters with attribution, materials in this book have not been previously published elsewhere in any format or language.

CONSERVATION AND PRESERVATION NOTES

The paper used in this publication meets the minimum requirements of American National Standard for Information Sciences–Permanence of Paper for Printed Material, ANSI Z39.48-1984.

Introduction

In this book I challenge the following conditions in the international industrial marketing field:

- The lack of theoretical development and the confusion it has created. I attempt to sort out this blurred situation of conflicting marketing theories, while at the same time introducing a new theoretical perspective to industrial marketing: the marketing economics approach.
- The lack of knowledge about how transnational corporations market industrial products outside Western industrialized countries, represented in this book by Southeast Asia. I use an institutional approach to address this specific deficiency.
  The combination of these two challenges involves a primary step to form a theory of international marketing of industrial products.
- The mainstream type of research methodology in industrial marketing, where logical deductive theories are tested by using large samples. This perspective is replaced by a theory-building case study approach, which is considered more relevant for the industrial marketing problems discussed in the book.

To meet these challenges I offer a new approach to industrial marketing–the marketing economics approach–by reporting a comprehensive study of industrial marketing strategies of transnational corporations in Southeast Asia. It merges different perspectives and theories into a powerful theory on international marketing of industrial products, mainly modern approaches to marketing, organization theory, and institutional economic theory. I also combine micro and macro approaches, which is rarely done in marketing and economics. In the book I present and illustrate this new framework by giving a detailed account of the experience of 13 West-European transnational corporations in industrial markets in Southeast Asia. I thus combine detailed empirical evidence with an advanced theoretical framework for analyzing international industrial marketing behavior.

My major aim is thus to provide an understanding of industrial marketing of international companies in this Asian region. A large part of the book describes how transnational corporations (TNCs) operating there tend to act and the sorts of decisions they make. The reader is given access to the industrial marketing experience of the TNCs which were studied. Many case studies are presented that follow a certain logic in elucidating how companies market their industrial products in Southeast Asia. Such case studies, if they are to be illuminating, must be set within a relevant theoretical framework. The framework that I develop is based on a transaction cost approach, considered a useful basic theory on how marketing activities are related to the economic environment, or more precisely on how industrial marketing is related to the vertical industrial system. The vertical aspect is not covered adequately in industrial marketing theory. Rather, market analysis normally focuses on horizontal market structures. I use Governance Forms theory in considering the institutional framework or context for vertical transactions. Marketing strategies are seen as being constrained by the governance form within which transactions take place.

Thus, I adopt a transaction cost approach to industrial marketing, analyzing in economic terms the strategic behavior of firms which operate in Southeast Asia. Transaction costs represent the driving force behind the economic organization of marketing in the area, pointing to factors of interest to strategic industrial marketing. Of overriding importance is the efficiency motive. The theory casts light on how transactional market failures come about and how such failures can be exploited in the building of protective barriers, especially those constructed through the creation of specific buyer/seller relationships. The marketing of industrial goods is thus very much a question of building and maintaining such linkages.

The matters I take up in the book represent in part a basic contribution to the theory of industrial marketing. The development of the theoretical framework involves two major steps. First, I give an account of how industrial products are marketed in less industrialized countries, and make an effort to develop a theory specifically pertaining to this. Second, in an attempt to develop an industrial marketing theory of more general character, I explore in depth the applicability of an institutional and a transaction cost approach within this context. I place particular emphasis on the international implications of the theory thus developed.

The matters taken up in the book can thus be considered highly relevant to industrial markets other than those studied in this project. The fresh perspectives the book presents and the development of new and innovative analytical techniques provide useful insights into industrial marketing. Throughout, I stress the importance of taking a broad approach to international marketing. It is vital that one not become so engrossed in how marketing has been conducted earlier that one loses sight of what the presence of a differing environment may imply. An institutional approach can be very helpful in analyzing the specific, contextually dependent conditions that can affect marketing. Both economic and noneconomic institutions are important in their effect on marketing based on transaction costs.

I provide the reader access to a broad fund of experience on the marketing of industrial goods in Southeast Asia through numerous examples of how different transnational corporations market a wide assortment of industrial products such as materials, components, equipment, spare parts, after-sales service, and projects. The case studies, together with the theoretical framework which contributes to an understanding of them, should be of interest not only to researchers, teachers, and students, but also to persons involved in management or in marketing operations in that geographical area of the world.

For the practitioner the book can serve as a rich source of marketing experience. Industrial marketing managers can consider carefully the experience of the companies represented in the book and ponder what is to be learned regarding their own situation. The executive should have no difficulty in drawing his/her own conclusions regarding the practical matters of individual interest.

Hans Jansson

Chapter One

The Marketing Economics Approach to Industrial Marketing in Less Industrialized Countries

A transnational corporation (TNC) in the industrial sector began marketing a package deal in Southeast Asia consisting of major equipment, knowledge, and service, as well as of maintenance, repair, and operation (MRO) items. This was the first effort by the TNC to market its package in a tropical country and the first test ever of the company's old and well-established know-how outside of Europe. The initial order received was from a fairly small Chinese firm that wanted to diversify into another branch of the industry. Problems arose almost immediately. The seller's technical know-how proved inadequate to the new and differing conditions prevailing in the area. Also, the equipment was unable to cope with the high humidity and hard tropical rains. In addition there were cultural problems, among these difficulties for the parties in understanding each other. As a result, the customer felt quite cheated. Only the fact that the Chinese customer hesitated in breaking off the relationship because of fear of "losing face" saved the day. Otherwise, the TNC might have been forced to stop doing business in the area before really getting started. A word-of-mouth agreement was reached to operate the equipment jointly. The seller also made considerable efforts to adapt the equipment and the operation of it to the tropical climate. The best engineers available were sent there from the European offices and managed to solve the problems. The TNC also stationed one of its men at the customer's plant to assist in operating the equipment. The final result was that the customer became the best manufacturer in its line of business in the home country and a very satisfied customer indeed. The TNC did lose a lot of money on the customer. This was necessary, however, for adapting the package deal to tropical conditions and for learning how to market it in a less industrialized country (LIC). The corporation tested the equipment and the technical know-how and built up market know-how at the same time that it created a reference project which potential customers could visit. The TNC was now ready to go on to establish itself in Southeast Asia (SEA) in a more substantial way.

The TNC took about five years to become firmly established in SEA, and in 1987/88, when a good level of sales had finally been reached, the turnover of MRO items had also stabilized at a profitable level to become the basis for the company's business in the area. The company is now well known within the market there, as shown for example by the many enquiries coming in from potential customers. The package deal has been gradually modified. The need to design technologically simpler products became obvious. A new version of the basic equipment was developed, to which different options could be added. This concept made it possible to adapt the equipment and price levels to the differing needs within the market, in turn making it possible to also expand to other sectors. In addition, the increased flexibility facilitated the deepening of relationships with many different customers, which are now initially offered simpler equipment which they can upgrade later as sales rise and sophistication increases.

Thus far the ASEAN (Association of Southeast Asian Nations) countries have been a virgin market for the TNC. No competitors are established or are even represented in the area. The considerable demand for MRO items which many of the customers have shown has gradually been discovered by certain international competitors. However, they market their products from offices in Europe and North America. Also, when customers are offered lower prices for such items by competitors, they tend to contact the TNC, which can then make appropriate adjustments. Prices do play a greater role now than earlier. At the start, when competition was almost completely lacking, technical matters were of major concern. During an initial period of three or four years, which involved all phases of the marketing process, it was essential that the needs of the market be met, to see to it that the equipment worked properly, to assist customers in operating it, to test product quality, and the like. The obvious rationale behind this was that, through helping the business of customers to thrive and expand, one increased the amount of MRO items and new equipment customers bought. No formal contracts for follow-up sales were required, or would have even been desirable, since they are not popular with the Chinese, who make up the majority of customers. If one insisted on such a contract, this could make it easier for other firms which did not sell on contract to compete. Nevertheless, whatever a TNC in such a position does, the risk from new competitors does increase over time, with customers feeling more and more secure about looking around for other suppliers.

This case illustrates certain industrial marketing problems with which TNCs are commonly faced in geographically and culturally distant markets. Such markets differ very much from the vast markets of North America, of such regional trading blocs as the European Community (EC), and of the highly industrialized countries generally. The markets themselves and their operating conditions, that is, the marketing institutions and the cultural and social values the markets reflect, differ, and the level of technology utilized by customers and by the regular suppliers is lower. In such remote markets the infrastructure also tends to be weak and the governments to be more involved in the control of specific business operations. This poses questions of how TNCs should best market their products there and of what marketing strategies should be employed.

The question of what industrial marketing strategies can work best under such conditions is considered in this book. Three basic elements of the problem are considered, as shown in Figure 1.1: the TNCs, the LICs, and the industrial marketing strategies which are appropriate, the major focus being on the latter. Thus, this is a book not on industrial marketing strategies generally, but on those applying to transnational corporations in their operations in less industrialized countries. The focus in the first chapter is industrial marketing strategies.

FIGURE 1.1. The focus of the book.

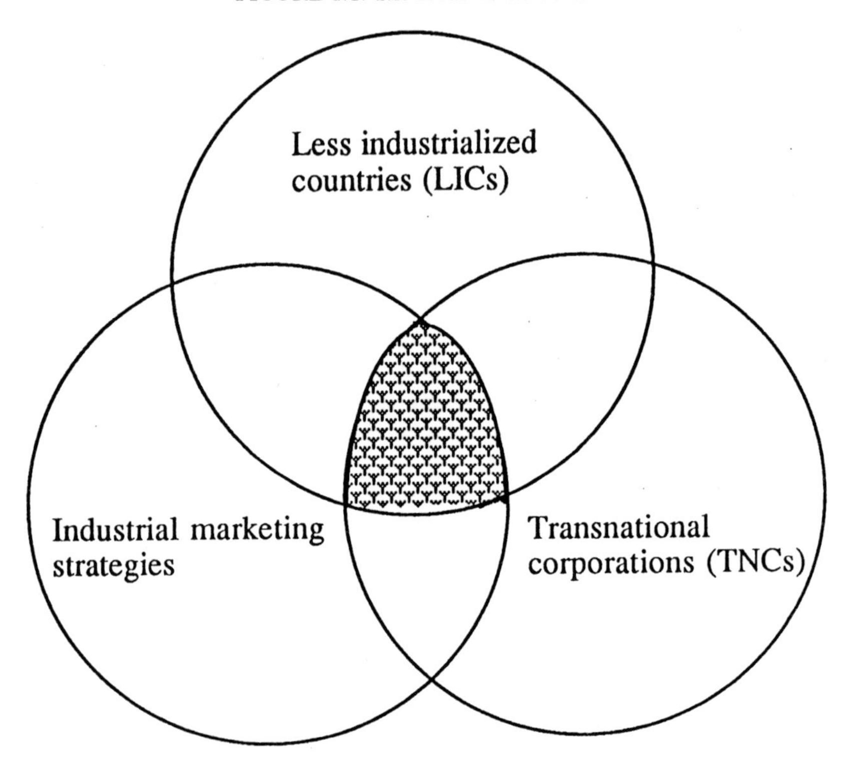

INDUSTRIAL MARKETING STRATEGIES

Studying industrial marketing means examining vertical relations in the economy. Firms act along a vertical value-added chain, some companies providing raw materials to steel works, for example, which in turn market certain of their products to car manufacturers, which in turn sell their end products to consumers. Internationally, there are marketing gaps to be bridged between those customers which are located in LICs and the production units, which are often located in industrialized countries (ICs). Sometimes such gaps are bridged through locating production units closer to the customer, or even within the local market. Whether or not this is done, transactions are necessary if the TNC is to learn of both the needs of customers and the competition present in these markets. Solutions which will satisfy customers' needs are bargained and later implemented. Resources (e.g., products, service, information, and know-how) flow between the seller and the buyer. Social interaction here is important to allow the transacting parties to affect each other in ways conducive to their overall goals. Such activities can be analyzed partly in terms of seller strategies, or of linkage strategies through which customers are selected and different channels to them, or linkages with them, are established. The competitive offer made to them consists of a mix of product features, service aspects, know-how, price, etc. The competitive strategy this involves, and the linkage strategy connected with it, are selected against the background of the needs of the customer and the offers made by competitors. The TNC's main objective here is to improve profitability and to enhance the company's long-term competitive position. This can be achieved through increasing efficiency, the most efficient company being the most competitive one. Efficiency is viewed in this book as involving a relative reduction in transaction costs. Efficiency as thus conceived is closely related to effectiveness as this is generally defined, that is, in terms of the attainment of predetermined goals. The more conventional conception of efficiency, on the other hand, concerns the relationship between inputs and outputs. In the present context efficiency and effectiveness are basically synonymous, or very close in meaning, the emphasis being on the organization's functioning in the attainment of a long-term competitive position. Allocative efficiency, which concerns resource allocation within the economy as a whole, or the question of whether production is efficient in relation to consumer demand, is viewed as separate from this.[1] Efficiency is thus considered, in this book, at the organizational and not the societal level, as involving the reduction or minimizing of transaction costs.

This conception of industrial marketing emphasizes the external organization of transactions as compared with their internal organization, and views the efficient organizing of transactions as representing a vital competitive strength. Nevertheless, it is not enough for the customer interface alone to be efficient. The latter must also be combined with other types of transactions for determining whether certain functions are best performed within or outside the firm. One frequent question of this sort, for example, concerns whether the TNC should have a sales organization of its own, or should let a separate sales company, whose services are hired and which serves as an agent, take care of sales.[2] That specific question will not be considered here in detail. Instead, the focus is on industrial marketing as such, whatever the role of external organizations may be.

A large part of the book, after a presentation of the theoretical framework, describes how TNCs in SEA tend to act and the sorts of decisions they make. Possible specific implications for the industrial-marketing management of companies new to the geographical area are discussed in the concluding chapter of the book. Concrete advice is provided regarding how industrial products can best be marketed in SEA. This is based on analyzing in economic terms the strategic behavior of firms which operate in the area. A transaction cost theory developed in an early section of the book is shown to be highly useful. The theory casts light on how transactional market failures come about and how such failures can be exploited in the building of protective barriers. The major barriers involved are those constructed through the creation of specific linkages, which in turn can be seen as representing assets.

A transaction cost approach to industrial marketing and purchasing is adopted. However, transaction cost theory in its conventional form does not suffice for providing an explanation and understanding of marketing behavior in this context. In its conventional form, transaction cost theory is an economic theory of neoclassical origin dealing with general economic behavior in a very abstract way. Neither the element of central interest in business analysis, namely the company, nor such functionally important areas as that of marketing are taken up specifically in most versions of the theory, which consider the firm as merely representing one of several possible forms of contract, one involving principal-agent contract or contract founded on property rights. Nevertheless, in one version of transaction cost theory, the Markets and Hierarchy theory, or the Governance Forms theory as it is called in this book (Williamson 1975, 1979, 1981, 1985),[3] the conception of the firm as an organization is more fully developed. Although this theory too is partly based on neoclassical economic theory, organizational aspects play an important role. It views the firm as an institution characterized by a particular organizational structure. Although the theory is not ascribed to in its entirety, it is of particular interest as it relates internal transactions to the organizing of external transactions with other institutions.

PURPOSE OF THE STUDY

The major aim of the book is to provide an understanding of industrial marketing in transnational corporations in Southeast Asia. Many case studies are presented similar to the one above. They follow a certain logic in elucidating how companies market their industrial products in the area. Such case studies, if they are to be illuminating, must be set within a relevant theoretical framework. A major task of this book is to develop such a framework based on a transaction cost approach. It also makes use of important elements of Governance Forms theory, considered a useful basic theory on how marketing activities are related to the economic environment, or more precisely on how industrial marketing is related to the vertical industrial system. The vertical aspect is not covered adequately in industrial marketing theory. Rather, market analysis normally focuses on horizontal market structures in accordance with, for example, oligopoly theory and industrial organization theory in their traditional form, which give little attention to vertical aspects such as bilateral oligopoly and vertical integration. A perspective emphasizing such aspects is developed in this book. Governance Forms theory is used here in considering the institutional framework or context for vertical transactions. Marketing strategies are seen as being constrained by the governance form within which transactions take place. In Chapter Three the governance forms involved are discussed in detail.

Governance Forms theory in its original form is a general theory of economic institutions. However, it does not address the strategic behavior of the individual business firm. A major aim of the book is to bring this general transaction cost theory down to the level of the organization, to functional areas such as marketing and purchasing. This involves a reconceptualization of the whole approach to be taken to industrial marketing, to viewing it as mainly an economic rather than a social discipline.

The matters taken up in the book thus represent in part a basic contribution to the theory of industrial marketing. The development of the theoretical framework involves two major steps. First, an account is given of how industrial products are marketed and purchased in less industrialized countries, and efforts are made to develop a theory specifically pertaining to this. Second, in an attempt to develop an industrial marketing theory of more general character, the applicability of an institutional and a transaction cost approach within this context is explored in depth. Particular emphasis is placed on the international implications of the theory thus developed. An initial discussion of this is provided in Chapter Two, with a more thorough discussion in Chapters Three and Four.

Chapter Three concerns, more specifically, institutional aspects of the theory, and Chapter Four depicts models of industrial marketing strategy. Chapters Five, Six, and Seven, in turn, present numerous case studies of how industrial products are marketed in Southeast Asia. The case studies, together with the theoretical framework which contributes to an understanding of them, should be of interest not only to researchers, teachers, and students, but also to persons involved in management or in marketing operations in that geographical area of the world. Questions of how to act in a practical way there are dealt with in Chapter Eight. The matters taken up in that chapter can also be considered highly relevant to other industrial markets. The fresh perspectives the book presents will ideally provide new and useful insights into industrial marketing and lead to the development of new and innovative analytical techniques. Above all, however, the reader with genuine interest is given access to the industrial marketing experience in Southeast Asia of the TNCs which were studied. Numerous examples are offered of how these TNCs market their many types of products there.

A major message of the book is that, although there are many commonalities in industrial marketing, differences abound. How best to market an industrial product in detail varies greatly, depending on the product and on a multitude of factors in the markets. Practitioners may find some parts of the book to be more relevant than others, depending upon how close they are to their individual experience. If the reader is lacking in previous experience with industrial marketing, the book can serve as a rich source of marketing experience. Industrial marketing managers can consider carefully the experience of the companies represented in the book and ponder what is to be learned regarding their own situation. It is the aim that with the help of the general conclusions presented in Chapter Eight, executives should have no difficulty in drawing their own conclusions regarding practical matters of individual interest.

INDUSTRIAL MARKETING THEORY

All approaches to industrial marketing acknowledge basic differences between the marketing of industrial and consumer products. A first major difference concerns market structure. Analyzing the market structures for the two types of products indicates the market concentration to be greater for industrial products, in that the number of buyers is much smaller. The 80/20 rule, stating that 20 percent of the customers are responsible for 80 percent of the seller's turnover, may well be valid here.

A second major difference in the marketing of the two types of products is that the interrelationships within the system are closer in the case of industrial products, especially in the vertical dimension. That dimension concerns interrelationships pertaining to various stages of production, normally involving firms and industries operating in areas relevant to a given line of production. Horizontal relationships, on the other hand, concern the market itself and reflect the effect of (consumer) incomes (Jansson 1982). The vertical dependencies can be characterized as involving derived demand, which is more fluctuating and volatile than demand in the horizontal dimension. Industrial market products are more interrelated, or joint, in existing demand, particularly in the vertical but also in the horizontal dimension, than are consumer products. Dependencies are represented in industrial marketing theory in various ways, being expressed, for example, in terms of networks which involve both vertical and horizontal dimensions. The buyer-seller relationships are closer and the distribution channels more direct in industrial than in consumer marketing.

A third major difference between the two types of markets is that for industrial markets interrelationships between parties are more stable and long-term (e.g., Mattsson 1975; Håkansson and Wootz

1975; Håkansson 1982; Hallén 1980; Jansson 1982). Various studies have shown that the industrial structure in many countries is rather rigid, with a marked preponderance of large companies. For industrial markets, geographical concentration tends to be greater, the numbers of customers to be less, and the individual customer to be greater in size than for consumer markets, the competition often being oligopolistic instead of monopolistic.[4] Similarly, demand in industrial markets tends to be more inelastic than in the marketing of consumer products. On the basis of the differences mentioned thus far, one can conclude that industrial markets are more specific and can be analyzed in only a marginal way in terms of traditional economic market models, a matter which will be discussed in some detail in the book.

A fourth dissimilarity between the two markets has to do with the type of demand. Industrial needs are generally more complex and sophisticated, in particular from an engineering point of view, than are consumer needs. The awareness of these needs and means of fulfilling them tend to be organized in a more professional way in industrial markets. The buyers there are firms, not individuals. For all of these reasons, sellers are confronted with a different demand situation there than in consumer markets, as seen, for example, in the buying process, which is far more complex in the case of the professional buyer.

THE MICRO-MARKETING APPROACH

A major impression to be gained from the literature is that most industrial marketing research is not founded on any theory of its own. Micro-marketing theory, for example, is used in its conventional form in connection with the marketing of both industrial and consumer products. The term *micro-marketing* is taken from El-Ansary (1983). Micro-marketing theory involves theories of product-brand management, pricing, promotion, physical distribution management, marketing research, financial aspects of marketing, and marketing program productivity. Micro-marketing theory can be contrasted with macro-marketing theory, a theory of consumer behavior, and with different theories concerning distribution channels. A perspective on marketing based on micro-marketing theory will be termed a *micro-marketing approach*. This approach will be compared with two other major approaches to industrial marketing: an inter-organizational approach and a marketing economics approach.[5] The micro-marketing approach emphasizes marketing management, which largely dominates research on marketing in North America. It is based on the principle that there are no major differences between the marketing of consumer products and industrial products.

The Marketing-Mix Approach

There are two different branches of a micro-marketing approach: the marketing-mix approach and the business-to-business marketing approach. The first of these treats ostensibly quite differing marketing perspectives as if they were essentially similar, through focusing on what is universal in the aims of marketing, namely consideration of the individual or organizational needs of the customer, needs which can only be met through a marketing mix. Kotler (1984, 4), for example, gives the following definition of marketing: "Marketing is a social process by which individuals and groups obtain what they need and want through creating and exchanging products and value with others." Similarly, Reeder, Brierty, and Reeder (1987, 8) write: "industrial marketing is, then, human activity directed toward satisfying wants and needs of organizations through the exchange process." In its mainstream research on consumer marketing, the marketing-mix approach is strongly anchored in the behavioral sciences.[6] A similarly strong link with behavioral sciences can be seen in the marketing management approach which is reflected in most textbooks on marketing published since the 1960s (e.g., Howard 1957; McCarthy 1960; Kotler 1967). A generally accepted and widely used definition of marketing management is that of Kotler (1984, 14): "Marketing management is the analysis, planning, implementation, and control of programs designed to create, build, and maintain beneficial exchanges and relationships with target markets for the purpose of achieving organizational objectives."

This marketing-mix approach can also clearly be seen in various textbooks on industrial marketing management which strongly reflect research and practice in the industrial field (e.g., Ames and

Hlavacek 1984; Chisnall 1989; Corey 1976; Hill, Alexander, and Cross 1975; Reeder, Brierty, and Reeder 1987; Haas 1989). Industrial marketing textbooks of this genre tend to identify major differences between consumer products and industrial products. However, one can contend that such differences are basically no greater than the differences in marketing strategy and marketing mix encountered when marketers of consumer products operate in different environments and have to change their strategies accordingly. Traditional micro-marketing models thus take no basically different approach to industrial than to consumer marketing.

The Business-to-Business Marketing Approach

One implication of the fewer and larger customers in industrial markets is the importance of the buyer/seller relationship. Although some researchers have previously emphasized the importance of this relationship (e.g., Webster 1979; Jackson 1985), it has only recently been given special attention. This attention has been so great that the study of buyer/seller relationships is sometimes viewed as quite a new field in the highly specialized American research on marketing, where it has been termed business-to-business marketing. Within this field there is "a uniquely strong emphasis on the unique elements of the buyer-seller connection" according to Bingham and Raffield (1990, ix), who see the reasons for this in such considerations as the following:

> a couple of years ago this text would have been called "industrial marketing management," but smokestack industries are on the decline, and course and text titles are changing to reflect the broader range of enterprises that rightfully fall under the umbrella of business marketing. Almost every available product or service is either aimed at business users or has a business marketing facet. Financial services, company car fleets, construction cranes, trade magazines, industrial lubricants, corporate jets, convention services: the list is vast. (vii)

Thus, the number and the types of business users have increased and changed. One result of this has been the inclusion of service marketing within the area of industrial marketing generally. The further implications of this "new look" or special emphasis, however, have not been developed much, in particular on a theoretical level. Such a neglect or "blind spot" is evident within the business-to-business marketing approach. The approach to marketing there has been little more than to explicitly focus on the buyer-seller relationship, or to consider it to be basically a market segmentation problem, involving the need of a special marketing mix, since customers are committed to few main vendors rather than being able to easily shift purchases from one supplier to another (Jackson 1985). Thus, this industrial marketing literature is largely practically oriented, not based on any special industrial marketing theory.

Although there is a general theory of marketing exchanges (Bagozzi 1975, 1979), it has not been employed appreciably in the industrial marketing field. Perhaps this is based on its formality, its bias toward consumer products, and its focus on social exchange and influence at the personal level.[7]

THE INTER-ORGANIZATIONAL APPROACH

The inter-organizational approach to industrial marketing takes account of the major differences discussed above between the marketing of industrial products and consumer products. Just as in business-to-business marketing and relationships marketing, the main focus is on buyer/seller relationships, on the business transactions taking place directly between the manufacturer and the user. In industrial markets, such buyer/seller relationships or linkages are often well-established and of long-term duration. They are complex, with contacts between the companies occurring on several different levels, particularly when complex products are involved. Particular emphasis is placed upon analyzing these relationships. This involves considering such questions as what the relationships consist of, how they are created, and how they are maintained. Such questions boil down to one main issue: how linkages between industrial firms are *organized*. Industrial marketing is viewed, according to this approach, as an inter-organizational matter, and it is regarded as fruitful to treat both the marketing and the buying behavior of firms as organizational issues. This makes it possible to apply the same theory of inter-organizational character to both types of activities (Jansson 1986). The two are seen as opposite sides of the same coin, which together represent the buyer/seller relationship. Interaction is viewed as taking place between the two parties to this relationship. The result is a more coherent theory for the field of industrial marketing generally. This is fundamentally different than relationships marketing or business-to-business marketing, in which traditional marketing concepts and tools are applied, albeit to direct marketing through relationships. As Bingham and Raffield (1990, ix) observe regarding the business-to-business marketing approach: "This approach brings forward the factor that most clearly shapes business-to-business marketing: the unique needs and processes of the business buyer. It is the buyer, after all, that distinguishes consumer marketing from business marketing. . . ." In this approach the particular focus is on the buyer and on employing common marketing tools within the framework of established relationships. Linkages are not viewed primarily as involving interactions from both sides. Indeed, an inter-organizational approach to industrial marketing does not mean that traditional marketing issues are absent or ignored. They are only put in their correct organizational context.

The most comprehensive interactionist model is presented in Håkansson (1982). It takes as its point of departure the following empirical observations:

- The parties, that is, the buyer and the seller, are both active on the market.
- The relationship between the buyer and the seller tends to be long-term, intimate, and characterized by a complex pattern of interactions within each firm and between the two firms.
- The linkages between the firms are institutionalized in the form of role-sets, in accordance with which the parties are expected to behave.

The last two points are especially relevant in connection with recurrent transactions between companies involving materials and components. They likewise apply to various patterns of transactions involving investment goods. The model considers four groups of variables:

- Variables depicting the parties of the relationship both as organizations and as individuals.
- Variables describing the elements of the interaction and the interaction process.
- Variables characterizing the environment within which interactions take place.
- Variables describing the atmosphere influencing the interaction and influenced by it.

In the interaction process, the model distinguishes short-term episodes from more long-term aspects of the process as expressed, for example, in the concepts of adaptation and institutionalization. Such a distinction is also made in this book, as transactions are viewed as conceptually separate from linkages.

The inter-organizational approach is strongly sociological in its orientation (see, e.g., Håkansson and Östberg 1975; Håkansson 1982; Hägg and Johanson 1982; Johanson and Mattsson 1987a, 1987b). Originally, it was applied to buyer-seller relationships between European firms (Hallén and Johanson 1989), but it has also been applied to buyer-seller relationships in India and Southeast Asia (e.g., Jansson 1982, 1986, 1988). Relationships or linkages are seen as developing through interactions within industrial networks. The main concepts of this approach are illustrated in Figure 1.2, which is taken from Johanson and Mattsson (1987, 38). The interactionist model involved concerns industrial marketing and purchasing behavior. Both the various exchange processes through which relationships develop and the adaptive processes that take place between parties in the continuous evolvement of relationships, for example through product modifications, delivery systems, financial systems and other routines, are analyzed. The central goal is to establish linkages or bonds which create dependencies between the parties involved. The more intensive the processes of exchange, the stronger the reasons for adapting to each other and not replacing the other party. A mutual orientation is created which results in a preparedness to interact in a dyad. A mutual knowledge of and respect for each other's interests is established. Mutuality is thus clearly characteristic of relationships within industrial markets. Such mutuality is largely shaped by social exchange processes.

FIGURE 1.2. Relationships and interaction in industrial markets.

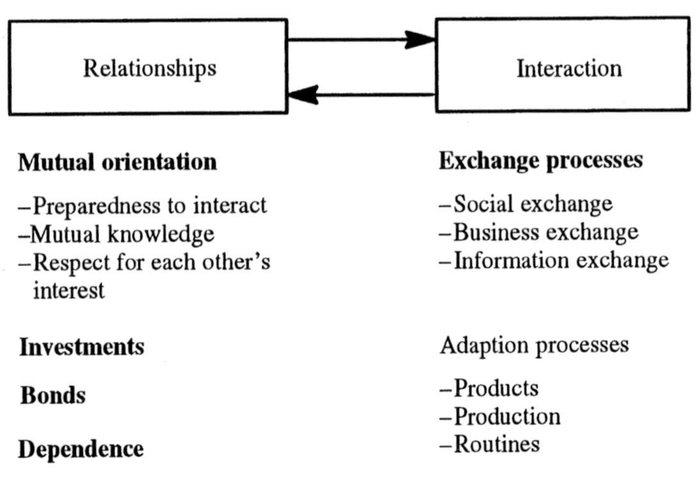

Source: Johanson & Mattsson (1987, p. 38)

Accordingly, means of transacting other than those of traditional economic exchange play an important role in industrial markets. Different forms of investment in relationships shape the future behavior of the parties involved, since they affect the parties' access to resources. It is less expensive to trade with companies with which one has linkages than with those with which one has none. Competitive position is likewise affected by such linkages. Aiming at establishing linkages involves realization of the fact that resources are heterogeneous. The same line of reasoning can be applied both to dyadic relationships and to networks of relationships. In the latter case, in particular, it can have a strong systemic impact on the marketing and purchasing possibilities of the firm involved. The firm's interactions within and without can be strongly affected by the access it has to resources from different types of networks. For example, how a seller interacts with its buyers depends partly on how the two-party relationships involved are connected to other, indirect linkages within the network or networks to which the seller has access, influencing how the seller relates to its suppliers. Within a given network different macro positions (relating to the whole network) and different micro positions (relating to given dyads) can be distinguished (Johanson and Mattsson 1987a).

Most research based on an inter-organizational approach focuses on dyadic relationships. Recently, the frame of reference has been extended to include networks of relationships. Nevertheless, the term *network* serves here mainly as a metaphor for the environment of the dyads. Although firms are indeed surrounded by networks, marketing does not tend to be analyzed at a network level. When networks are considered in this context, it is generally the interaction between two parties in a network and not that between several parties which is examined. Accordingly, the inter-organizational approach tends to mainly regard only interaction which involves single links and not sets of links.

THE MARKETING ECONOMICS APPROACH

A micro-marketing approach builds implicitly on certain premises of microeconomic theory, for example that consumers and sellers are rational, and that markets are in equilibrium. The emphasis is on discrete transactions rather than on relations. This can be seen clearly in the pricing models which the approach employs. If price is considered at all, it is usually treated as a separate matter, as a strategy separate from strategies of other types. In this way the economics of marketing is relegated within that approach to a "back seat," with emphasis placed instead on more general social scientific aspects. In this book I attempt to remedy this bias and to clearly emphasize the economic foundations of marketing, interpreting marketing behavior in economic terms. To this end, transaction cost theory is here introduced into marketing area. This allows marketing behavior to be considered calculative in nature. Greater emphasis is placed too on pricing, which is seen as an integrated part of marketing and as the main component in competition. This provides a clear perspective on the marketing of industrial products. It does not mean that traditional microeconomic theory is reintroduced into marketing theory. Rather, the approach is one based largely on institutional economic theory.

A marketing economics approach is closely related to an interactionist approach, with marketing viewed as an organizational issue in both approaches. However, the former is based on economic theories of organization rather than on inter-organizational theories. The focus on dyads is similar for both approaches. However, the network metaphor is not used to characterize environmental variables within which interaction takes place in the marketing economics approach. These variables have been replaced within that context by institutions. Vertical linkages are focused on and institutions are analyzed more specifically as governance forms. These matters will be discussed in detail.

The major points of the marketing-economics approach that differ from the sociological inter-organizational approach are as follows:

1. The mutual orientation of the inter-organizational approach is replaced in the marketing-economics approach by a competitive orientation, in which conflict rather than cooperation is emphasized, in line with the transaction cost approach on which the marketing-economics approach is based. It is the vertical dimension in competition which is stressed, a dimension which is of particular importance in industrial marketing. Thus the parties to a transaction are seen as competing for the resources involved. This differs considerably from what applies to the marketing of consumer products, where horizontal aspects of competition prevail and transactions can be considered to reflect conditions on the horizontal dimension within the market. In the marketing economics approach, therefore, competition on the vertical dimension, in particular, is emphasized, at the expense of cooperation.

   The emphasis on competition does not mean that cooperation is completely neglected, but rather that it is viewed primarily as a means for resolving conflicts. This is based on a view of man as more prone to conflict than ready for cooperation. A major consequence of this fundamental premise is that relationships are viewed as less harmonious and parties as more prepared to break off relationships when competitive situations change than would otherwise be the case.
2. Behavior is viewed as basically calculative and not based on trust as a natural human condition. The book takes as its point

of departure in this respect a kind of principle that "in the beginning there were conflicts," that economic behavior has its roots in market institutions and not in the family or in some other institution of which one can say that "in the beginning there was cooperation and trust." This can be contrasted with the emphasis in the inter-organizational approach on long-term relationships which appear, on the surface at least, to be of cooperative character.
3. The establishment of bonds or linkages which create dependencies is analyzed in economic and not in social terms.
4. Exchange and adaptation processes are seen as primarily steered or governed by considerations of efficiency rather than of power, where examining power relations is more typical of a sociological than a purely economic approach. The problems involved in attempts to view power and efficiency jointly are illustrated in research on distribution channels (e.g., Stern and Reve 1980; Reve and Stern 1985).
5. Systemic influences on industrial marketing are viewed as coming from markets and market-like institutions rather than from networks.

Besides being a highly fruitful approach in Southeast Asia, the theoretical perspective the marketing-economics approach provides also has other advantages:

- The theory is more uniform than that of an inter-organizational approach. One of the main pitfalls of the network approach is avoided, namely the confusion or confounding of analytic levels. A fundamental ingredient of the inter-organizational approach is social exchange theory (Blau 1964, 1968). This theory tends often to be employed in the same way in analyzing interpersonal as in analyzing inter-organizational exchange. As Blau (1987) observed, there is the danger here of transferring basic concepts from the micro-level to higher levels of analysis.[8] The same critique can be directed against the consideration of power in this context. Such critique is evident in Emerson's approach to power (e.g., Emerson 1962; Cook and Emerson 1984), which focuses on individuals and not on organizations. In addition, power theory as employed in the inter-organizational approach is not related explicitly to economic perspectives, making it ambiguous as a theoretical foundation to the theory.

- The theory behind a marketing-economics approach is also more general than that behind an inter-organizational approach. Thus, the theory makes it possible to relate industrial marketing behavior at the firm level to other types of economic behavior, as well as to noneconomic behavior of an institutional character. The transaction cost approach connected with it also makes it possible to study the relation between industrial marketing and its organization within the same theoretical framework. This has not been done in research based on an inter-organizational approach (see, e.g., Håkansson and Östberg 1975; Håkansson 1982). Thus, organization theory and inter-organization theory have not been merged. A transaction cost approach provides such a possibility since external and internal transactions can be compared within a single theory. Thus, the consideration both of industrial marketing and purchasing behavior and of the conditions which affect such behavior or under which it takes place can be integrated within one and the same theory.

All in all the marketing economics approach, as compared with the other two approaches mentioned, provides a simpler theory which concentrates on the essentials of industrial marketing. Considerably fewer concepts are needed than in the other two approaches.

AN INSTITUTIONAL APPROACH TO INDUSTRIAL MARKETING

Industrial marketing is conceived in this book as involving the organizing of transactions so as to bridge gaps between needs and solutions in an industrial system. Linkages are distinguished from transactions here. A transaction is a separate exchange of, for example, products or of know-how (information). A linkage or relationship is taken to mean the framework within which recurrent transactions occur and which influences and is influenced by individual transactions (Jansson 1982, 4). A transaction is thus the basic unit, whereas a linkage is an aggregated set of transactions representing a composite concept. Linkages and relationships (relations) are used synonymously here. Four major types of aggregated transactions or linkages represent the bridges between needs and solutions, namely product, information (know-how), social, and monetary (financial) linkages.

This conception of marketing represents the adaptation to industrial marketing of the notion of marketing gaps or marketing separations found between production and consumption (McInnes 1964).[9] This position tends toward a functional approach: "Marketing is viewed from the standpoint of activities or functions. The entire field is broken down into a limited number of economic services or functions such as buying, selling, transportation, and so on" (Lichtenthal and Beik 1984, 144).[10]

Industrial marketing should be characterized, however, in a more precise way. A functional approach is too narrow for our purposes, since it yields too limited a view of the environment within which marketing takes place. The study of marketing in LICs requires a broader perspective to allow marketing to be placed in its institutional context. The concept of comparative marketing focuses directly on this aspect (e.g., Boddewyn 1981; Kaynak 1984). This can also be seen in textbooks on international marketing (e.g., Cundiff and Hilger 1988). The systems approach to marketing is broader and relates closely to this approach.[11] Carman (1980) presents a systems/exchange paradigm, in terms of which different approaches to marketing can be analyzed. In applying this paradigm to research on vertically integrated and quasi-integrated systems, he concludes that it is only Williamson's (1975) work which deals with all the constructs in the systems/exchange paradigm. Williamson's analysis is described as having resulted in unique contributions regarding the goals of the system, environmental constraints, internal control systems, relative power of each actor, transaction costs, and the level of information available. In discussing different types of exchange Carman concludes:

> With regard to closing the systems defined as the discipline of marketing, an analysis of gifts, marriage, religion, and politics

> led to the generalization that marketing exchanges: (1) should be conducted for the primary purpose of resource allocation; (2) require ingredients of freedom and trust for the parties; (3) exist in an environment with sufficient property rights for the parties to exchange use rights. "Resources" here include a very broad category of "goods." By these criteria, politics, marriage, and religion were excluded from our discipline. (Carman 1980, p. 32)

As is implied above, this book is distinguished from most treatises on industrial marketing through the emphasis it places on economic aspects.

TRANSACTION COSTS

Transaction costs can be defined as the costs for the organizing of transactions between units in an industrial system for the purpose of identifying and influencing industrial wants and needs and providing solutions to them.

Transaction costs are thus a way of interpreting marketing costs. They are a mental construct used to explain economic behavior,[12] consisting mainly of the costs needed to get information about and/or influence (create) a need, find a solution to the need, and implement the solution, that is, the costs for the exchange of products, information, and money, as well as for social exchange. According to Lichtenthal and Beik (1984), G. S. Downing gives a comprehensive definition of marketing from a systems standpoint, which also resembles the above definition:

> Marketing is the process by which the firm keeps surveillance of its markets, detects and evaluates forces for change, and feeds this back as inputs into the firm, thus generating throughout all the strategic decision points in the firm new strategies and action, with new or adjusted output behaviour designed to defend against goal-obstacles or to exploit opportunities. Customers and competitors respond; their responses are observed, evaluated, and fed as inputs back into the firm; and again throughout all the strategic decision points in the

> firm, strategic adjustment may again be actuated. (Lichtenthal and Beik 1984, 151)

To this definition, which mainly expresses how information costs and decisions are connected, we would add bargaining and enforcement costs. Our definition of transaction costs is largely based on Coase's (1937, 390) definition, according to which transaction costs are the cost of discovering the appropriate prices. These costs include the costs of information, measurement and negotiation. This definition is based on the assumption that the needs are "out there," and that it is simply a matter of discovering them. Our marketing-oriented definition adds to this the possibility of the supplier creating needs or influencing them. Individual tastes and preferences are not viewed as exogenous factors as they are in orthodox economics. They are considered to be formed by the parties continually, as well as by their social environment and by the general environment, for example, by non-economic institutions. However, Coase's emphasis on the costs of information, measurement, and negotiation is important to build on. Therefore, transaction costs are seen as not only incurred in the discovering and influencing of prices, which is mainly related to the informational aspect, but also in bargaining about prices and fixing them, together with controlling that the prices agreed upon are enforced.

A similar definition of transaction costs is found in Hennart (1982, chapter 2). Transaction costs are defined there as the external costs of the economic system. In a world of perfect information, perfect enforcement, and no bargaining costs, as in neoclassical economic theory, there are no external economies. Hennart's approach is based primarily on McManus (1975), who views enforcement costs as the main external cost:

> I define the cost of enforcement to be the resource cost incurred to detect violations of behaviour constraints. When we count our change, weigh meat, punch time clocks, inspect a used car, or supervise labourers, we are incurring costs to enforce behaviour constraints by monitoring or measuring the activity of another individual. (336)

Whether the market or a hierarchy is chosen for enforcement purposes depends, in McManus' view, on which mode of organization appears most likely to maximize the consumption possibilities of the individuals in society. The least inefficient-appearing method of allocation is chosen. The three main transaction costs are thus information costs, bargaining (negotiation) costs, and enforcement costs. Measurement costs are an aspect of all three, since they relate to the problem of quantified costs in general.

These various costs are incurred because of divergences from a perfectly competitive market model.[13] Whenever there is a lack of perfect knowledge in the market, information costs arise. This also makes it impossible to retain the assumption of profit maximization by producers and of maximization of utility by consumers. An abandonment of the premise of atomistic competition means that large sellers and buyers may have an individual influence on prices and yet they do not act completely independently. Costs for negotiating and influencing prices are incurred, these being particularly high due to the fact that there are many heterogeneous products on the market and not simply one homogeneous product. Enforcement costs are also incurred. Resources are needed for non-price competition as well, making the resource allocation process rather complicated. If the premise of free entry to and exit from every product market is relaxed, this further complicates the allocation process and also affects transaction costs. The assumptions inherent in a perfect competition model in which inputs are rewarded in direct accordance with their contribution to productivity may or may not be assumed to be valid for factor markets. If these and the other assumptions mentioned above are relaxed, the optimalities they involve disappear and information costs, enforcement costs, and negotiation costs come about in factor markets as well. As Coase showed in 1937, the relaxation of these various assumptions makes room for the existence of the firm in economics, allowing its existence to be explained in terms of economic theory. The firm is viewed as a more efficient contract than are markets in reducing transaction costs. Markets thus become internalized and prices become replaced by directives, for example, to increase the efficiency of the labor force. As Cheung (1983, 16) states:

> I do not claim that price determination is the only transaction cost which matters in all choices of the forms of contracts or organizations. But with regard to the central point made by Coase, my own investigation supports his view: under private property rights, any movement toward the contractual delegation of use rights results mainly from the constraints of pricing costs.

Williamson and Ouchi (1981, 388) describe transaction costs in the following way: "More generally the analysis of transaction costs focuses attention on *alternative means of contracting*. A preoccupation with technology and steady-state production expenses gives way to the study of the *comparative costs of planning, adapting, and monitoring task completion.*" The main task to be completed is defined here as that of satisfying a particular need. Planning, adapting, and monitoring are the means for achieving this task. In the process, costs for information (e.g., decision-making), bargaining, and the enforcing of decisions are incurred.

Transaction costs are incurred by the parties to a transaction. Such costs are defined from the points of view of the business units and are thus private and not social costs.[14] Transaction costs have to do both with the exchange process as such and with its outcome. Efficiency in transacting and in the distribution of results are related. The size of the pie and how it is distributed are interconnected. Rules for distribution are determined in the transaction process. Viewing matters in this way is a departure from neoclassical economic theory, in terms of which the distribution of the result is automatically determined by market forces. In this equilibrium, inputs are paid for in accordance with their contribution to the result. An equity rule tends to be followed such that a constant ratio is maintained between the inputs and the gains of each party.[15] The distribution of the results in turn gives the parties different resource conditions for future transactions.

Conditions under which the transaction process can change can also create new conditions for future transactions, introducing a dynamic aspect both to the buyer/seller relationship and to the environment in which it exists. Among other things it can affect competitive positions and environmental constraints. This aspect will be developed primarily in Chapter Seven, in connection with the possibility for a firm to obtain first mover advantages through a fundamental transformation process.

This transaction process is itself constrained by outside factors, for instance, by laws and rules which regulate transactions. A TNC may not be interested, for example, in transacting with customers in a country in which neither laws (e.g., defining ownership rights to certain vital assets) nor customs secure a particular outcome under given conditions.

PRODUCTION COSTS

As conceived here, transaction costs are the major marketing costs for bridging and organizing gaps between sellers and buyers. Costs for the transfer of solutions are included but not those for producing them. Production costs for manufacturing a product to fulfill a particular need are thus not included. However, certain costs closely related to manufacturing are defined as transaction costs. These are the information, bargaining, and enforcement costs concerning where to locate different stages of production in relation to the customer. Manufacturing costs are defined as the costs for producing an item, that is, the costs of the factory (plant) and of production within it. These costs constitute constraints on a company, which cannot select a particular strategy on the basis of transaction costs alone. Even if transaction costs could be lowered by moving production closer to local customers, high manufacturing costs brought on by not being able to take advantage of economies of scale could inhibit this. It is thus important to study how different aspects of a company's manufacturing process constrain transaction costs, for example, economies of scale and the production technology. In addition, there is a connection in the other direction, in terms of the impact of the transaction frequency and the degree of transaction-specific investment on the costs of transport, tariffs, and multiplant economies (Nicholas 1986, 74). There is also the impact of other institutional factors. A production investment could be brought on by nonmarket factors, such as government stipulations.

Costs normally defined as production costs in the calculus of the one party, certain inventory and transport costs, for example, are part of the costs for transferring a solution, and are thus defined as transaction costs. These are costs for exchanging products.

Both transaction costs and production costs influence vertical industrial systems. For example, high efficiency in production as achieved in many industries through economies of scale tends to create oligopolistic market structures, which can increase bargaining costs between firms. On the other hand, the high degree of uncertainty often experienced by TNCs when operating in developing countries can lead to their taking on "in-house" production of certain inputs they would otherwise purchase. Low internal demand can then result in low production efficiency due to insufficient use of economies of scale.

INSTITUTIONS

Transaction costs can arise in transactions either between different economic units, or within a particular market or a firm, with economic institutions involved in each case. The meaning of economic and of noneconomic institutions is elucidated in Figure 1.3. According to institutional economics, economics is not a universal science and economic models cannot be used out of context. The neoclassical tradition, in contrast, is founded on the principle of universality, a principle which institutional economists repudiate. Most economic research carried on outside the mainstream neoclassical tradition appears to be of an institutional character. A wide variety of economic theories are involved, as widely differing as Williamson's Governance Forms and Myrdal's (1968a) Social System Theory. Williamson works very much within the confines of mainstream economics, in terms of which institutions are conceived as governance forms. They are means of exchange which differ from those of the neoclassical market. They include both firms and other market forms. Transaction costs are for Williamson, the mental constructs needed to explain these other institutional forms. A still broader approach is that of Myrdal (1968a), who puts economics into its societal context, considering economic factors to be one of the various types of factors in a social system.

The approach taken in this book is related to both Myrdal's and Williamson's institutional approaches. The emphasis on the impor-

FIGURE 1.3. Economic institutions and noneconomic institutions.

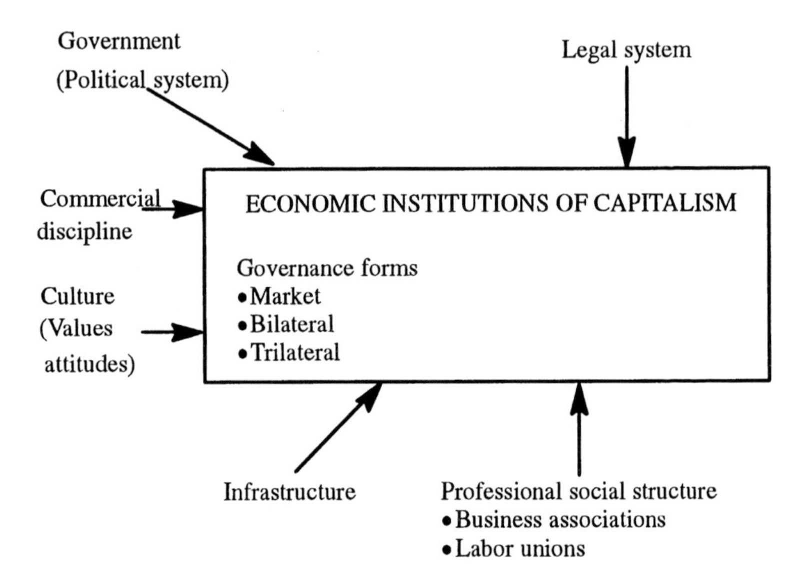

tance of the wider societal context for understanding economic behavior in developing countries follows Myrdal (1968a). Three major noneconomic factors are stressed: attitudes, institutions, and the state. The first two can be related in a general way to culture, whereas the third plays a critical role in the planning and evolution of official policies. Various forms of economic organizing behavior that will be referred to are taken from Williamson (1985). His theory considers only the legal role of the state. The various economic institutions of interest are complemented by the consideration of different noneconomic institutions as well, as shown in Figure 1.3. The mechanisms by which these institutions are related to each other and to industrial marketing are analyzed throughout the book. The approach taken is based on the assumption that economic behavior follows an efficiency logic expressed mainly in these various institutions or governance forms. The behavior involved is also considered within a broader institutional context, mainly that of culture and of the state. Thus, although economic behavior is emphasized, it is assumed to be affected or constrained by society at large.[16] This standpoint places the approach taken close to that of various other institutional approaches to marketing, in particular that adopted in certain research on marketing systems, although Joy and Ross (1989) indicate that the approach there tends to be closer to that of Myrdal's theory than of Williamson's theory. There has noticeably been no attempt in the literature to combine aspects of Myrdal's and Williamson's theories.

A consequence of the institutional approach taken here is that, in analyzing industrial marketing, both systemic functions and institutional aspects are considered. Dyadic exchanges are seen as being supported by economic, legal, and other institutions, the major purpose of which is to reduce transaction costs as much as possible in order to achieve an efficient exchange of resources for a particular industrial need. Characteristics of the transactions determine within which economic institutions they are organized. Conversely, economic institutions vary in their ability to organize different types of transactions, thus establishing boundaries.

Economic institutions in turn are confined or also supported by other, noneconomic institutional frameworks, for example, those of the culture, the legal system, and the state. There, for instance, are common values and attitudes within a society, expressed in various noneconomic institutions, which either facilitate or obstruct the formation of certain economic institutions which are calculative in nature and oriented toward efficiency. The legal system is one institution involved here:

> Market exchange requires a combination of both state and customary institutions. For any developed system of commodity exchange there must be a legal system enscribing and protecting rights to individual or corporate property. There must be a body of contract law with criteria for distinguishing between voluntary and involuntary transfers of goods and services, and courts to adjudicate in such matters. In addition, however, the evolution of law is not simply a matter of legislative construction; a great deal of law grows out of custom and precedent. Property and contract law are not exceptions. Consequently, the existence of property and exchange is tied up with a number of legal and other institutions (Hodgson 1988, 150).

Costs external to economic institutions for creating laws, common values, and other infrastructure in order to facilitate efficient transactions are normally not included in transaction cost analysis. They constitute basic conditions which are prerequisite to efficient economic transactions.[17] The basic institutions involved cannot simply be taken for granted in LICs and left out of the analysis. The external costs such institutions might otherwise be expected to take on may have to be internalized by the parties themselves. In this book, noneconomic institutions are analyzed in terms of how the transaction costs of the economic institutions of interest are affected by them. Deficiencies in the basic institutional set-up, such as in the legal system, increase transaction costs, producing among other things greater uncertainty. The costs of establishing and running such institutions are normally borne by the state in ICs, where they are seen as providing a public service. In LICs, on the other hand, they may need to be internalized by individual economic agents, becoming private goods or property.

Thus, transaction cost theory can be used to relate noneconomic to economic institutions. This is in accordance with the main purpose of the book: to explore possibilities of finding a common theory that can explain both industrial marketing behavior itself and how various other phenomena are related to it.

LESS INDUSTRIALIZED COUNTRIES

Due to the large differences between ICs and LICs in their institutional set-up an institutional approach to industrial marketing is particularly fruitful here. The efficiency with which an economic institution can be moved or transplanted to a new setting or environment is partly dependent on the inherent characteristics of the institution. Thus, the basic form and set-up of the economic institutions in the developing country, its market structure for example, are partly determined outside the developing country. TNCs are organized in specific ways which are partly transferred to the new setting. One important question then is how well such an "imported" structure is matched to the institutions native to the developing country. Consideration of dynamic aspects is important too. As the country develops its industrial structure, the factors inherent in various imperfections in the market can change in character. Thus, various institutions within the market there become more advanced and function better from the TNC's standpoint than at the initial stages of development, making it easier for the company to shift certain internal transactions into the market.

An assumption made in this book is thus that industrial marketing strategies in LICs cannot be studied without considering the context within which they operate. Unhappily enough, most business problems in developing countries tend to be studied with the implicit assumption that the economic institutions of various countries are so similar that markets work in basically the same way in such differing countries as Indonesia and the U.S. The approach of this book, in contrast, is institutional or contextual, with emphasis placed upon the specific characteristics of the institutions within which economic actions are embedded. Different marketing strategies, to be sure, are examined within the basic context of the economic institutions of capitalism. However, the extent to which strategies employed in one country can be transferred to other countries is dependent upon the extent to which the economic institutions of the one country are similar to those of the other. This is necessary in efforts to transfer strategies and the like between ICs and LICs. This basic situation is illustrated in Figure 1.3. Transactions take place through any of these economic institutions or governance forms. The effects are dependent upon how efficiently the various economic institutions function. This in turn is affected by the noneconomic institutional framework, in which the market institutions involved are embedded, specifically government, the legal system, the culture, and professional structures. Accordingly it is seen as necessary that careful account be taken of the norms, laws, and rules that affect the transaction process involved. In contrast, such neoinstitutional theories as transaction cost theory treat the economic framework as one of basically universal character.

Market structures in developing countries can differ from those in industrialized countries in many ways. The functioning of market mechanisms in such countries tends to be impeded by various factors. Capital is usually scarce and concentrated to the few. The lack of infrastructure and defects or weaknesses in the information transmittal mechanisms accentuate the difficulties. Power concentrations and dual market structures may exist, which can create barriers to entry and enable certain groups to exercise leverage in controlling the entire system or other actors within it.

THE ASEAN COUNTRIES

The five ASEAN countries of interest here (Indonesia, Malaysia, the Philippines, Thailand, and Singapore) have shown remarkable rates of economic growth during the 1970s and 1980s. Some key figures for these countries are given in Table 1.1. Singapore is a newly industrialized country (NIC), and Malaysia is on the brink of qualifying for that epithet. This difference can be seen in Table 1.1, as can the gap in industrial development between these two countries and the other three. Thailand's high growth during the 1980s is gradually narrowing the gap. However, the difference is still rather large. Indonesia and the Philippines are clearly farther from such a level of economic development. The major role of exports in their growth is a common characteristic of these countries, with manufacturing taking the leading role as an engine of growth and a vehicle for transforming the economy. At first, each of the countries, with the exception of Singapore, exported primary products such as raw materials, agricultural products, and forestry products. Manufactured products were then gradually added to these. During the period of transition, they practiced an import-substitution policy. Their success can largely be explained by their having found a good mixture of these traditional development strategies: primary products exports, import-substitution, and industrial specialization. This latter specialization has been accomplished by low labor costs and other favorable terms they could offer in the industrial export zones. Such a mixture of policies takes advantage of their rich raw material base, a large domestic market, and their opportunities to participate in an international division of labor. The latter has involved their opening their economies to international competition. The contribution of manufacturing to GDP and to employment is thus rather high in the four countries to which this applies. Singapore, which has an economic structure highly skewed toward manufacturing and the service industries, rates particularly high on this point.

Another major characteristic applying to all five countries is that

TABLE 1.1. Performance figures for five ASEAN countries.

	Indone-sia	Malay-sia	Philip-pines	Singa-pore	Thai-land
Area('000 km²)	1,948	330.4	300	0.625	514
Population					
Total	184.6	17.4	64.9	2.7	55.6
Urban pop. as % of total	26	35	41	100	17
Workforce					
% manufacturing	8.3	17.3	20.6	28.5	10
Social					
Telephones ('000)	1,253	1,247	788	1,200	1,100
Cars	1.32m	1.55m	376,646	238,984	1.15m
Trucks and commercial	1.50m	3.8m	598,209	108,477	337,000
Length of railways (km)	4,432	1,644	539	38	3,735
Length of paved highway (km)	46,255	39,000	n.a.	2,810	44,868
Production					
GDP at market prices (US dollars) 1988	64.15	24.54	39.19	24.5	56.1
Per capita income	403.7	1,875	667.35	8,162	1,038
GDP, real growth '88 (%)	3.59	8.7	6.4	11	11
Manufacturing as % GDP	13.93	25.6	25.09	30.1	24.4
Public Expenditure					
Education as % budget	7.74	18.1	13.13	20.8	17
Foreign Trade					
Merchandise exports (US dollars) 1988	19,218	20,220	7,074	32,714	15,842
% manufacturers	48.2	54.7	64.6	25.6	19
% food and farm products	10	n.a.	21.8	4.5	34
% metals and minerals	23	12.8	11.1	0.9	0.8
Merchandise imports					
% plant, capital equipment	23.5	45.7	23.01	43.4	39.6
Public debt-service ratio (%)	n.a.	11.8	n.a.	0.4	13

Source: Asia 1990 Yearbook, pp. 6-9

participation in world manufacturing has taken place primarily through the presence of TNCs. Traditionally, the main investors came from Japan, Europe, and the U.S. However, from about the end of the 1980s onward, Far Eastern firms, largely from Taiwan but also from Korea, have been increasing their investments in the area sharply. The ASEAN countries have not developed any large corporations of their own as Korea did. The domestic manufacturing industry consists largely of small-scale and medium-sized firms run and controlled by the Chinese business community. This biased firm structure may be a major deterrent to fast and sustained growth. For one thing, modern technology has not spread far outside the foreign TNCs. There has also been much short-term speculative activity in trade, finance, and property, as well as strong efforts to gain commercial advantage through political patronage.

As can be seen from the figures in Table 1.1, the infrastructure tends generally to be good in Singapore and Malaysia and to be weak in Indonesia and the Philippines, with Thailand somewhere in between. However, with the high growth rates in Malaysia and Thailand the infrastructure becomes more and more overburdened. Indonesia and the Philippines, in particular, are handicapped by large populations and large geographical distances between their many islands.

The role of the state has been crucial for the high growth rate. These countries are highly managed market economies. In a way they resemble the modern transnational corporation in their joint application of authority and market control of the economy. Private fiat is replaced by public fiat. The countries vary in their policies toward direct foreign investment. The Philippines and Indonesia are especially known for having strong governmental regulation and control. Together with Malaysia they have a strong preference for joint ventures, with locals as majority shareholders, particularly for operations in domestic markets. The cultural influence on how governments and domestic companies are managed in Southeast Asia (SEA) is very strong. As Pye (1985) has shown, the political systems vary a lot between the countries reflective of strong cultural traditions. In the private business field similarities between the countries are much greater due to the dominance of the Chinese. The colonial impact is still felt in the ASEAN countries. The British influence is noticeable in Singapore and Malaysia, the Dutch influence in Indonesia, and the U.S./Spanish influence in the Philippines.

TRANSNATIONAL CORPORATIONS

Strategic industrial marketing as practiced by TNCs differs from that practiced by domestic firms, among other things because of the TNC's superiority in technology. Moreover, the marketing carried on by the local subsidiary of a TNC is backed and influenced by the very large parent organization outside the host country, from which many of the strategies may originate. Local strategies are often transferred then from the group and coordinated with the global strategies of the corporations.

THE TNCS STUDIED

A brief summary of the subsidiaries studied and of the groups to which they belong is presented in a number of different tables in this section. Certain basic information about the organization, sales, employees, production, and products of the various groups is given in Table 1.2.

The industrial products which these TNCs market in Southeast Asia are specified further in Table 1.3. The products in question are classified into seven different groups. The outputs for group one are raw materials, for group two processed materials (e.g., steel, aluminum, hard metals and packaging material), for group three castings and forgings, for group four components and subassemblies, for group five minor equipment, for group six major equipment, for group seven Maintenance, Repair, and Operating (MRO) items (e.g., spare parts). The products span a large number of different industries, as shown in Table 1.4.

As can be seen in Table 1.5, most TNCs (exclusive of Brunei) are well-represented in all the five major ASEAN countries. A major reason for this is that the activities in SEA are a "natural" part of the worldwide operations of these companies. Nine of them are large multinationals active in most parts of the world (Table 1.2). However, one of these (Paving systems) carries on only minor activities within ASEAN. The remaining four likewise have a large degree of internationalization but are smaller than these major TNCs. One of the four companies (Wear and Tear) is a rather small

TABLE 1.2. Profiles in 1990 of the Transnational Corporations studied.

NAME OF THE GROUP	Pacmat	Wear & Tear	Paving syst.	Food equip.	Tele-com	Indpow	Conmine
GROUP ORGANIZATION							
No. of business areas/ divisions	F	–/2	8/–	7/–	6/–	8/65	3/7
No. of companies	–	10	300	–	150	1300	50
East Asia	–	1	9	6	13	–	10
Southeast Asia	4	1	5	3	8	–	3
SALES							
Total (in US$billion)	5	0.1	1.6	5.6	9	25	3
Europe (%)	55	–	80	53	60	57	54
East Asia (%)	3	–	9[1]	7	6	15[2]	9
Southeast Asia (%)	–	1.5	–	2	–	–	–
EMPLOYEES							
Total (thousands)	12	1	25	21	70	141	21
Europe (%)	–	–	95	70	80	65	56
East Asia (%)	–	1.5	0.2	3.5	4[1]	9	12
Southeast Asia (%)	–	1.5	0.07	1.7	–	–	–
PRODUCTION							
East Asia	Yes	No	Yes	Yes	Yes	Yes	Yes
Southeast Asia	Yes	No	Yes	No	Yes	No	No

NAME OF THE GROUP	Tooltec	Weld-prod	Explo	Ind-comp	Drives	Specmat
GROUP ORGANIZATION						
No. of business areas/ divisions	6/–	4/–	F	5/–	4/–	F
No. of companies	–	48	–	180	40	55
East Asia	10	5	–	17	3	8
Southeast Asia	4	3	2	9	2	4
SALES						
Total (in US$billion)	3.6	1	–	5.6	0.5	0.5
Europe (%)	60	60	–	63	–	–
East Asia (%)	–	–	–	–	–	–
Southeast Asia (%)	–	–	–	–	–	–
EMPLOYEES						
Total (thousands)	26	8	–	49	4	3.5
Europe (%)	70	42	–	75	83	80
East Asia (%)	3.5	5	–	1.6	3	7
Southeast Asia (%)	0.6	1	–	1	0.5	3
PRODUCTION						
East Asia	Yes	No	Yes	Yes	No	No
Southeast Asia	Yes	No	Yes	Yes	No	No

1 = Includes the whole of Asia; 2 = + Australia/New Zealand; F= Functional organization

TABLE 1.3. Main groups of industrial products sold in Southeast Asia by the TNCs studied.

PRODUCT GROUP	TRANSNATIONAL CORPORATION
1. Raw materials (RM)	
2. Processed materials (PE)	Pacmat, Tooltec, Specmat
3. Castings and forgings (CF)	
4. Components and subassemblies (COM)	Indpow, Wear and Tear, Food Equipment, Indcomp
5. Minor equipment (MIE)	Indpow, Weldprod, Food Equipment, Tooltec, Conmine, Drives
6. Major equipment (MJE)	Indpow, Telecom, Pacmat, Paving Systems, Conmine, Food Equipment, Weldprod
7. Maintenance, Repair and Operating items (MRO)	Indpow, Explo, Wear and Tear, Food Equipment, Tooltec, Conmine, Indcomp

but highly internationalized TNC with a very small operation in SEA. This TNC is controlled by a holding company with a portfolio of very diverse companies.

All the TNCs studied are manufacturing companies which sell their products to other producers. As shown in Tables 1.3 and 1.4 most of these 13 TNCs offer a wide variety of different industrial products to industries of various types within ASEAN. However, a small number of them focus on only a few major products. The marketing strategies in selling these products are taken up extensively in Chapters Four through Seven.

The TNCs considered have gradually established themselves in SEA. Some started quite early, but most of the growth took place from about the end of the 1970s onward (see Jansson 1989). The organization of the TNCs varies considerably. The global and local organizations of the various groups is analyzed in Jansson (1992).

Compared with selling, manufacturing is only a relatively minor activity for the TNCs within the ASEAN countries. Production there is nevertheless important for five of the firms, and of much lesser importance for four of the others. The remainder of the firms have no

TABLE 1.4. Main customer industries.

INFRASTRUCTURE	
Construction	Conmine, Indpow, Explo, Wear and Tear
Roads, railways etc.	Indpow, Conmine, Explo, Paving Systems, Weldprod
Telecommunications	Telecom
BASIC INDUSTRY	
Power	Indpow, Conmine, Food Equipment
Steel	Food equipment, Drives
Shipping, off-shore	Indpow, Food Equipment, Drives, Weldprod, Conmine
Oil	Tooltec, Drives
Mining	Conmine, Wear and Tear, Drives
Petrochemicals	Food Equipment, Tooltec, Drives
ENGINEERING INDUSTRY	
Machinery	Tooltec, Indpow, Specmat
Electronics	Indpow, Conmine, Specmat
Packaging	Pacmat, Food Equipment
General	Conmine, Indpow, Indcomp, Tooltec, Specmat
FOOD INDUSTRY	Food Equipment, Drives
TEXTILE INDUSTRY	Indcomp, Tooltec
CERAMIC INDUSTRY	Wear and Tear
FOREST INDUSTRY	Tooltec

production whatever in the area. With a few exceptions, the typical production problems of LICs in SEA, mainly due to the import-substitution policies in practice, are not particularly germane. This is reflected in the purchasing strategies of the TNCs. Supplier development, for instance, has only been forced upon two of the TNCs. Most of the TNCs are allowed to import what they need rather freely.

This is only a brief introduction to the TNCs which were studied. Various operations of these companies will be illustrated and analyzed in the chapters that follow. The reader will learn more about them as many of their activities are discussed. The purpose, however, is not to provide a complete account of the operations of each and every one of them. Rather, common emerging patterns among them are described. In addition, only those companies which best repre-

TABLE 1.5. The representation in Southeast Asia of the Transnational Corporations studied.

	ASEAN Singapore	Malaysia	Thailand	Indonesia	Philippines	Rest of Asia
Pacmat	PC(RO)*	SC	SC*	A	SC	SA+EA
Wear and Tear	SC+*	A	A			EA
Paving systems[D]	SC+*	A	A	A	A	PA
Food Equipment	RO*;SC	SC*	SC*	A	A	
Telecom	RO*;A	SC[3]*;PC	SC[4]	A	A	
Indpow[87]	RO*;SC*	PC;SC*	PC;SC*	A	PC	
Conmine	SC(RO)*	SC*	A	D	SC	
Tooltec	RC+*	SC*	SC	D	SC	EA
Weldprod	SC(RO)*	SC	A	PC;A	A	EA
Explo[87]	A	PC(RO)*	A	A	PC	
Indcomp	RO+*;SC[3]; PC	SC,PC	SC	D	SC	PA
Drives	SC*	SC	A	A	A	
Specmat	RO*;SC	SC	SC	D	D	PA

Abbreviations:

PA = The Pacific-Asia basin except USA; SA = South Asia; EA = East Asia;
A = Agent; SC = Sales company; PC = Production company;
TrB = Trading company; D = Distributor; RC = Regional company; RO = Regional office

* = Companies interviewed
+ = The regional organization covers a larger area than the ASEAN, as shown in the right hand column
' = The regional company only has a mandate for a few countries in this region, i.e. Burma, Sri Lanka and Bangla Desh.
3,4 = The number of companies
D = Only the regional organization of the division studied is represented here
87 = This regional organization is from 1987, since the TNC was only studied between 1984 and 1987

sent the theoretical framework matters of interest are selected and included in the discussion in any given chapter. The distribution of the activities of the TNCs studied over various chapters is shown in Table 1.6.

OUTLINE OF THE BOOK

The book consists of four major sections. First, in Chapter Two the research methodology is described. This is followed by a theoretical section including a chapter on the institutional framework (Chapter Three) and a chapter on industrial marketing strategies (Chapter Four). The third section is empirical and consists of three chapters that describe industrial marketing strategies in Southeast Asia (Chapters Five, Six, and Seven). In a fourth and final section the implications for industrial marketing management of the material and ideas presented are discussed.

Since the main purpose of the book is to present a theoretical framework for explaining industrial marketing activities of TNCs in LICs, the sequence of chapters will follow that theory-building logic. Each chapter will be concerned at least in part with this theoretical framework.[18] The aim of the book is not the presentation of individual case studies as such. Those presented provide instead the empirical base for the development of the theoretical framework. The book thus synthesizes the evidence gleaned from the cases and organized around the major theoretical topics. Under each topic included in a chapter an attempted synthesis of the case material presented is provided. Appropriate examples are drawn from the cases illustrating specific aspects of the topic.

An overview of the TNCs dealt with in the various chapters is given in Table 1.6. Depending on how much information about the companies in question is provided, the latter are divided into two groups. Most of the illustrations of points in Chapters Five and Six are taken from the primary cases Food Equipment, Conmine, and Tooltec. Illustrations taken from the secondary cases are only used when they are found to complement these three major cases, specifically Pacmat, Wear and Tear, and Drives in Chapter Five, Weldprod in Chapters Five and Seven, Indcomp and Specmat in Chapter Six, and Paving systems in Chapter Seven.

TABLE 1.6. TNCs illustrated in the book.

TNC	CHAPTER NO.		
	5	6	7
Primary cases			
Food Equipment	X	X	X
Conmine	X	X	X
Tooltec		X	X
Secondary cases			
Pacmat	X		
Wear and Tear	X		
Weldprod	X		X
Indcomp		X	
Paving Systems			X
Telecom			X
Indpow			X
Drives	X		
Specmat		X	

The marketing of projects is an activity undertaken mainly by TNCs other than those included among the primary cases. Two companies, whose activities are only marginally illustrated in Chapters Five and Six are taken up in greater detail in Chapter Seven: Paving systems and Weldprod. Examples from Telecom and Indpow are new to the chapter. Certain minor examples from the three major cases, Food equipment, Tooltec, and Conmine, are also taken up in that chapter.

CONCLUSIONS

This book on the industrial marketing strategies of transnational corporations in less industrialized countries takes industrial marketing theories as its major point of departure. However, the present state of such theories was found to be inadequate for describing and explaining industrial marketing activity in such countries. To this end, therefore, a transaction cost theory was developed. The theory involves an institutional focus and represents an institutional theory of industrial marketing. The elaboration of the theory can be seen as constituting a general theoretical contribution to the area. The methodology employed is discussed in Chapter Two.

Compared with the sociologically inspired inter-organizational approach to industrial marketing, the marketing-economics approach assumes more of a competitive than of a mutual orientation. Behavior is considered to be basically calculative and not based on trust as a natural human condition. In addition, the establishment of relationships is analyzed mainly in economic rather than social terms. Exchange and adaptation processes are seen as governed primarily by considerations of efficiency rather than of power. The result is a more explicit theory as regards both the premises and theoretical levels of analysis. After showing a number of different definitions of marketing to be of only limited use in providing an understanding of industrial marketing, a transaction cost perspective is utilized. The term "transaction cost" means the cost of organizing transactions between the various units in an industrial system with the purpose of identifying and influencing industrial needs and providing solutions to them. The three major transaction costs–those of information, bargaining, and enforcement–are presented as based on deviations from the model of a perfectly competitive market.

Industrial marketing as conducted by TNCs within LICs is usually organized in terms of sales subsidiaries which are backed and controlled by a large organization operating worldwide. Strategies often originate from somewhere within a given group and are transmitted to subsidiaries, being coordinated in terms of global strategies. Industrial marketing theories adapted and adaptable to the particular context of LICs are created through focusing on transaction costs and institutional or contextual aspects. The connections which exist between economic and noneconomic institutions are illuminated in this way. This perspective is applied to industrial marketing within the ASEAN countries.

This book is based on a study of the operations in the ASEAN countries of 13 European TNCs which were presented briefly in this chapter. The cases these firms represent are taken up again in Chapter Two from a methodological standpoint. In accordance with the conditions stipulated for the interviews, the case studies themselves and the informants remain anonymous. It is essential to distinguish here between the way in which the theories have been developed against the background of these cases and the way in which the cases are used in the book. The overall theoretical patterns are based on the entire data base of the cases in question, whereas only certain portions of the cases are used in this book, which illustrate certain interesting aspects of the theories.

NOTES

1. Allocative efficiency, one element of which is price efficiency (or productive efficiency), refers to the extent to which the allocation of resources within the economy meets the efficiency conditions of Pareto optimality. Another efficiency concept is that of dynamic efficiency, which takes account of more long-term impacts not considered in more traditional forms of welfare analysis, such as effects that may result from structural and technological changes in the economy.

2. This conception accords with the eclectic paradigm and with the internalization theory of the TNC, both of which serve to explain why TNCs are more efficient than local firms in catering to local markets (see Jansson 1992).

3. The implicit dichotomy found in the Markets and Hierarchies theory, which stems from Williamson (1975), is avoided by the designation Governance Forms theory employed here. The latter also seems a more adequate term in the present context, in which the focus is on governance forms rather than on the market/hierarchy divide, and the typology is more varied, with intermediate governance forms such as bilateral governance and trilateral governance being considered.

4. Chisnall (1989, 61-62), for instance, concludes after reviewing different investigations of market concentration in Great Britain:

> Earlier it was noted that industrial markets are often dominated by a relatively limited number of enterprises with aggregate sales accounting for the bulk of industry sales in specific markets, a trend accentuated over recent years by industrial mergers in many industries, e.g., brewing, ship-building, carpet production, electronic engineering, etc. Buyers in these large firms wield considerable power in purchasing a wide range of products and services. The 80/20 rule has increasingly become evident in industrial markets. For instance, over 80 per cent of the capital products turnover of the UK-owned electronics companies is accounted for by five indigenous

> manufacturers: GEC, Plessey, Racal, Ferranti, and ICL. Some industries, however, remain considerably fragmented, e.g., house-building, or road haulage. . . . Fewer sources of supply will be open to buyers of several types of products or services.

5. The distinction will be made between theory and approach. An approach signifies a perspective which forms a basis for research. Theories developed in this book are based on looking at the world from an economic or efficiency perspective. They are also based on viewing the world as consisting of institutions rather than of networks. The theories concerned are conceptual representations or models of the world based on such perspectives. They consist of concepts (or factors) and statements of relations between these. The attempt is made to explain these relations or connections between the constructs and to determine and describe their direction.

6. As expressed by Alderson (1957, 8):

> the version of marketing theory here may be regarded as an aspect of the general theory of human behaviour. It is related to economics since it is conceived with efficient correlations of means and ends. It is more allied to the broader social sciences in its conception of the social setting in which the individual or the operating group seeks to achieve their ends.

7. For an evaluation of this theory, see Ferrel and Perrachione (1983).

8. After spending many years trying to merge microsociology and macrosociology, Blau (1987) came to the conclusion that

> the two theories require different, if complementary, perspectives and approaches. One major reason is that the two involve incommensurate conceptual schemes. Basic concepts of microanalysis, such as reciprocity, obligation, network density, or multiplexity, are not relevant for macroanalysis because the latter does not dissect social interaction and role relations between individuals. At the same time, basic concepts of macroanalysis, such as heterogeneity, inequality, and the degree to which various social differences are related, are emergent properties of collectivities that cannot refer to individuals and thus are not appropriate for the study of role relations between individuals. Even when the same term is used, its meaning in micro- and macroanalysis is quite different. Imbalanced obligations and dependence in exchange relations give rise to interpersonal power, but this superordination is neither the same as nor the source of the power of concern in macrosociological analysis–like that of the corporation executive, the general, the dictator–which rests on resources and authority positions. Both involve dependence, but interpersonal dependence and power are confined to relatively narrow circles of persons with whom one has direct or, at least, indirect contact, whereas large resources enable a person to make very large numbers dependent and hence to exercise power over them without even indirect contact with them. (99-100)

9. See also Håkansson (1982), Hammarkvist, Håkansson, and Mattsson (1982), and Gadde (1980).

10. Different marketing functions are illustrated by one of the classifications found in Lichtenthal and Beik (1984), namely the one suggested by F. E. Clark and C. E. Clark in 1942:

> A. Functions of exchange: 1. Selling (demand creation bringing buyer and seller together). 2. Assembling (buying). B. Functions of physical supply: 3. Transportation (physical transfer). 4. Storage (hold and preserve goods). C. Facilitating functions: 5. Financing (capital). 6. Risk-taking (merchandise loss of value or physical damage). 7. Market information (interpretation of facts or new facts). 8. Standardization (establishing standards). The performance of these functions constitutes the process known as marketing. All functions constitute marketing. (cf. Lichtenthal and Beik 1984, 158)

11. The similarity between the institutional approach to industrial marketing and a marketing systems approach is illustrated by G. D. Downing's definition,, which is summarized in the following way by Lichtenthal and Beik (1984, 150):

> (1) Marketing as an aspect of greater society (a socioeconomic process); (2) marketing vis-à-vis the total activities of the firms; and (3) marketing as the functions or activities performed by marketing people. These approaches interact with one another, and each could be studied separately.

12. Transaction costs as a mental construct should not be confused with use of the term in other economic contexts, such as in cost accounting or in "reality," as, for example, costs in bank transactions.

13. The assumptions of perfect competition are the following: 1. All firms in an industry produce a homogeneous product. 2. Business firms and consumers possess perfect knowledge of relevant alternatives. 3. Firms endeavor to maximize their profits and consumers to maximize their utilities. 4. Competition is atomistic. 5. Entry and exit are free in every market (Cohen and Cyert 1965, 5-6). Neoclassical theory is not designed to explain or predict the behavior of real firms. The firm represents a theoretical link there, a mental construct, helping to explain price determination in markets.

14. Transactions are the basic unit of economic exchange within society and considered to express a rationale for acting in a non-neoclassical economy as that present in the real world, in which imperfections are germane. The goal is changed there from optimization such as the minimization of transaction costs to that of satisfying such objectives as reducing transaction costs to a certain level. It is not a question of striving for the most efficient governance form but for the least inefficient one.

15. Most transaction cost theory takes this premise as a starting point and goes on to show what inefficiencies result when this relation is not present (e.g., Alchian and Demsetz 1972; Cheung 1983).

16. This is a different institutional approach than the one normally found in marketing. I examine the ways in which marketing takes place within both eco-

nomic and noneconomic institutions. The more usual approach is to study marketing in different types of organizations, such as in profit and nonprofit organizations, or in retailing and in wholesaling.

17. Institutional theories are employed in organizational analysis, which is a sociologically based approach to institutions, one involving no universal theory of organizations. Rather, it is assumed that organizations should be studied in their societal context. At the same time, society is conceived as manifesting itself within an organization, for example in terms of myths. The organization existing within a firm is assumed to follow principles which differ from those of say a school or a political party. A model for organizations within different societal sectors is to be found in Meyer and Scott (1983). Just as in various institutional economic theories, values and culture play a prominent role in institutional organization theories. "Selznick distinguished between organizations as technically devised instruments, as mechanical and disposable tools, and organizations that have become institutionalized, becoming valued, natural communities concerned with own self-maintenance as ends in themselves" (Scott, 1987, 494). Institutional organization theories are largely based on a phenomenological approach and have been developed as a reaction to traditional organizational theories which are based on assumptions of rationality and efficiency. They can therefore not be used in this research project that has this latter foundation.

18. Yin (1984, 133) calls this an approach of "theory-building structures," something which is relevant to both explanatory and exploratory case studies.

Chapter Two

Methodology

THEORETICAL PERSPECTIVE

An institutional approach to research is based on the conviction that behavior and social reality are very much a function of the social framework and context in which they occur, and so there is no generally valid truth. Values and beliefs place important constraints upon reality but these can certainly change over time. It is assumed that the meaning of a given economic behavior varies with the economic environment. In contrast, much of the research on TNCs from an economic perspective thus far has been of a neoclassical character, based on the assumption of a universal theory, one that allows behavior to be freely generalized from one country to another. In consumer and industrial marketing research within a micro-marketing approach it is often taken for granted that a given set of marketing techniques can be used in about the same way in less industrialized countries (LICs) as in industrialized countries (ICs). When this is shown to be impossible, the line of reasoning tends to be that the marketing theory is correct but that it has been wrongly applied due to the inadequate development of the market institutions of the LIC.

It is not a simple matter, however, to transfer a theory for understanding the behavior of the firm, which has developed mainly in one particular type of environment, directly to another environment. Phenomena in a new and many-faceted environment can often best be understood if multiple dimensions are taken into account. There is little justification for examining situational effects on certain aspects of marketing behavior in the new environment and assuming that others are unaffected by conditions. Marketing behavior is related to many factors, of which the researcher has very inadequate knowledge. Consequently, it is often not possible at the outset to discriminate properly between important and unimportant factors. One should not assume that markets in LICs and ICs function in the same way. For example, local cultural and governmental policies may have a strong impact on what a company in a LIC does. Also, deficient infrastructure, a lack of development of basic industrial structure, and a highly unequal distribution of income may constrain company action. The company's possibilities for marketing its products or services may also be restricted by both its own basic strategy and its organizational structure. Considering the entirety of such effects may be more important than considering those emanating from a particular source. In view of this, we choose to select a few cases and study them in depth, considering a wide variety of different aspects, rather than to examine many cases and study just a few aspects.

TELEOLOGICAL EXPLANATIONS

The theoretical perspective we take in this book differs from that implied in most transaction-cost theories, with their strong dependence on neoclassical economic theory. According to Kogut (1985) and Robins (1987), serious methodological criticisms can be directed against transaction-cost theory in general for its neoclassical bias. These same criticisms apply to the use of neoclassical trade theory and general transaction-cost theory in analyzing TNCs.[1] One criticism concerns the implicit neoclassical claim that the minimization of transaction costs results in the selection of an optimal organization, something which is akin to the equilibrium of a perfect market. Such a conception ignores the imperfect state of the real world. When real-world conditions prevail, an optimal equilibrium is lacking, efficiency is not unequivocally determined, and it is not possible to provide a causal explanation for the evolution of a particular social and economic structure (Robins 1987, 71).[2] For our purposes, there are two major problems with the neoclassical view. First, the resulting theory is too general. Second, the types of explanations it provides are inadequate. The actual economic behavior of organizations is thus not adequately explained by the theory. Its premises are formulated with the aim of creating a general and ideal model of economic efficiency, one inadequate, however, for the specifics and complexities of the real world of organizations.

I depart from all such premises in endeavoring here to create a more realistic model of efficiency, one based on a transaction-cost perspective. Such a perspective avoids the danger of arriving at nothing more than tautological explanations. Neoclassical economic theory is a logical deductive theory, as the conclusions which it provides are derived from the assumptions made and these in turn are connected by a number of general laws. The validity of the theory can only be tested within the framework of its premises. Testing the validity of a theory empirically requires that some proposition which is deducible from its assumptions be tested by experimentation or observation. There is a risk that such a highly general theory becomes platitudinous and redundant.[3]

Regarding the type of explanation sought, neoclassical theory employs both causal and teleological explanations. Neither of these, however, are of much use in a context such as the present one. Teleological explanations of such generality as those neoclassical theories employ are expressive of the logical interrelationships within the theory as such and not of the economic actors. In neoclassical terms, individuals strive to maximize utilities and firms to maximize profits. However, such hypothetical characteristics are far from any real-life intentions in the present context. In the approach taken here, economic processes are seen as occurring through the motives and actions of actual men and women, and through the actions and functioning of organizations and institutions, not apart from these.[4]

Hence, I dissociate myself from the very general economic theories which neoclassical theories represent, where individual economic actors are assumed away. The outcome is, of course, a less general model. What is lost in generality, however, is gained in reality. My purpose is to construct a more specific theory to explain certain types of economic behavior. This view also repudiates functionalist explanations, with their analogies to evolutionary theory in biology and to other purposes not specifically intended by individuals, which suffer from the same weaknesses as discussed above. They are too general, preoccupied with the functions of the overall system instead of with the intentions of the parts or of the individuals therein. There is also the risk of tautological explanations.[5]

In this book I employ a governance-form type of transaction-cost theory to examine relationships between certain environmental conditions and industrial marketing behavior. My major premise is that an organization's prime aim or purpose is to become more efficient. Accordingly, I seek to explain the organizational behavior of TNCs in terms of intentions, conceiving the organizational behavior involved in terms of the governance of the economic activities of individuals so that it coincides with the demands of a competitive environment.[6] The industrial marketing theory presented here is a conceptual framework for analyzing economic exchange within and between organizations. I assume economic institutions to be basically rational in their intentions. The general economic logic which I develop is applied to specific organizations and the particular settings in which they operate.

THE RESEARCH STRATEGY

From the standpoint of this perspective, the case study method clearly represents the research methodology of choice. Yin (1984, 23) states:

> A case study is an empirical inquiry that: investigates a contemporary phenomenon within its real-life context; when the boundaries between phenomenon and context are not clearly evident; and in which multiple sources of evidence are used. This definition not only helps us to understand case studies, but also distinguishes them from the other research strategies that have been discussed. An experiment, for instance, deliberately divorces a phenomenon from its context, so that attention can be focused on a few variables (typically, the context is "controlled" by the laboratory environment). A history, by comparison, does deal with the entangled situation between phenomenon and context, but usually with noncontemporary events. Finally, surveys can try to deal with phenomenon and context, but their ability to investigate the context is extremely limited. The survey designer, for instance, constantly struggles

> to limit the number of variables to be analyzed (and hence the number of questions that can be asked), to fall safely within the number of respondents that can be surveyed.

Case studies are to be preferred, according to Yin (1984), when analytical generalization is the main purpose of research. "In analytical generalization, the investigator is striving to generalize a particular set of results to some broader theory" (p. 39). This comes closer to the methodology used in experiments than that used in surveys. In the latter case, research is based on statistical generalization, that is, generalizing from a sample to a population. Analytical generalization is achieved through a replication process, where the relevance of various theories is studied for several different cases. The aim of analytical generalization is not to validate theories, but to study their suitability in unfamiliar economic environments. The generalization criterion is thereby replaced by a flexibility criterion (Brunsson 1982; Bulmer 1979). Theoretical sensitivity is important (Glaser 1978). The method is not purely deductive since it does not involve validating theories through testing hypotheses. Neither is it purely inductive: theory is not developed entirely from or grounded in empirical data.[7] Such pure forms of generalization can be considered to be naive and to readily lead to bias. One can consider the most adequate strategy instead to be to remain aware of the fact that knowledge can indeed be generalized to a certain extent between different research situations.[8] Researchers interested in the analytical validation of matters mentioned here can take as the starting point for their research the findings presented in this book. From this perspective the specific conclusions presented here are to be viewed as suggestive rather than definitive. One can also attempt to formulate hypotheses on the basis of the theoretical framework developed. The constructs one employs must then, in turn, be operationalized in terms of the purpose one has.

I employ a comparative case-study or multiple case-study method in this book. Cases have been researched as wholes. Subdivisions of companies, such as various departments or divisions, have not been studied individually. My case study is thus holistic and not embedded (Yin 1984, 44-47). The cases are the local subsidiaries of transnational corporations in Southeast Asia. A holistic approach is fruitful in research on TNCs in developing countries. Even if industrial marketing is focused, other main activities of the TNCs have therefore been included. These are mainly establishment processes, industrial marketing and purchasing, production, international organization, government relations, and business cultures.

Analytical Abstraction and Generalization

The theoretical framework presented here has been developed through a combined process of abstraction and generalization. In abstracting, the researcher concentrates on certain facts and disregards others. The more phenomena disregarded, the more abstract a theory becomes. At very high levels of abstraction it may be hard to recognize reality at all. In generalizing, the researcher expands a study from one or a few cases to many. As already noted, analytical generalization and not statistical generalization is employed.[9]

The practical approach taken in the work represented in this book was as follows. Before leaving for SEA, I developed a preliminary theoretical framework on the basis of research I had conducted earlier, primarily in India, and on a study of the literature. I also interviewed key informants at the headquarters of the TNCs to which the selected case companies belonged. The purpose here was to gain basic insight into the matters to be studied, become familiar with certain relevant-appearing sources of information, and obtain permission to study the subsidiaries of these TNCs in SEA. The theoretical framework was rather broad, since the aim was to gain an understanding of the major activities of the companies in question in the area, and at the same time was very asymmetric and imperfect. For certain activities which had been studied earlier, such as production and purchasing, there was a well-structured theoretical framework, one that had been largely developed in the Indian study. For others, such as marketing, the questions were based on more preliminary theories and it was not known how relevant these theories would be for researching industrial marketing in SEA. Furthermore, there was no relevant theory regarding local and regional organizations of TNCs in far-off markets, or regarding cultural and government influences on industrial marketing there. For these matters, questions were formulated in a very general way. What was primarily lacking was an overall theory for relating these more specific theories to each other and to the economic environment. The idea, already at the start, was to use transaction cost theory to this end. However, at that point it was not known how and how well this could be accomplished.

Thus, there was evidence which supported taking a certain construct validity of the approach in some of the theoretical areas but not in others. The establishing of the external validity of findings began already in the first field study, on the basis of recurrent replications of various findings as the cases unfolded. The fit of findings with theory was shown to be acceptable for those parts of the theoretical framework which were already best developed (concerning purchase and production). A careful inspection of the Southeast Asian data collected during the field study confirmed this. Regarding the theoretical areas that were less well developed, evidence for the relevance of certain factors was found to be stronger in the empirical material than that for the relevance of others. Specific patterns of factors could also be observed, to which explanations were sought. The interpretation was refined through further study of the literature. Gradually, a more equal theoretical coverage of the different areas under research–marketing, purchasing, production, organization and cultural influence–was achieved. Government influence proved to be not as important as expected. The pattern found for this in the field study was judged to be enough for the present purposes and no further refinement seemed called for, neither through a study of literature nor through a gathering of more data. At the same time, the theoretical integration of the findings in the various problem areas had not yet been established. In particular, relations of the overall findings to factors in the economic environment had not been dealt with adequately.

A case study base was developed. Case-presentation drafts were prepared for 17 cases, an additional company not being included here due to their failure to cooperate in this respect. The case-presentation drafts were then sent for review to one key informant in each of the companies, usually the managing director. Some of these were returned with only marginal corrections or comments. Most of them, however, were commented on in detail. Some of the informants refused to send the draft back, declaring that it could only be reviewed in an extended dialogue with the researcher. The reliability of the study, and as a consequence its validity as well, appeared to be at stake. Therefore a new trip to visit the case companies was made half a year later, mainly to review the data collected during the first field study. However, this also provided an excellent opportunity to follow-up the earlier study. The previous long boom in the area was now rapidly turning into a slump. For some of the companies this was close to representing a negative turning point in their developments.

Two more case companies were added to the study and persons there were interviewed during that second trip. The status of the project was also evaluated. Since the case companies studied thus far were all subsidiaries of very large TNCs, it was considered important to compare their activities with those of the subsidiaries of smaller TNCs. As noted below, it is important in a study like this to examine cases which vary greatly in activity, industry, size, etc. Four such subsidiaries were selected and persons there were interviewed. However, because of insufficient response only two were retained for further study.

Altogether, data for 19 cases had now been collected in the manner planned. The internal reviews of the two new cases could be obtained by correspondence. During the sessions with the key respondents the data were divided into classified and nonclassified information. The purpose here was to clarify what information could be published openly and what information could not. In the process most of the cases were also reviewed by persons at the TNC headquarters in Europe.

The individual portions of the theoretical framework were elaborated further and the consistency among them was increased through the consideration of organization theory and of theories of business strategy. However, the various business theories selected or developed took no account of questions of the business environment, above all of matters of culture and government. In one article by the author, the attempt was made to combine organization theory, transaction cost theory, and theories of multinational firms, with the aim of providing an understanding of the organization of the TNCs. This led to insight into how transaction cost theory could be incorporated into the theoretical framework, both to increase the consistency between different parts and to relate them to business environmental aspects.

One aspect of the data base in particular contributed to a realization of how transaction cost theory could be used for this purpose. Practically all the subsidiaries studied were faced with strong price competition in Southeast Asia, both in high-quality markets and when they were involved in long-term buyer-seller relationships. This was contrary to what had been found in industrial marketing research in Europe, where price was of much lesser importance. This insight made it especially clear that an efficiency approach to marketing could be more relevant in SEA and raised serious questions concerning the more usual, sociologically oriented approach to industrial marketing employed quite generally for analyzing conditions in Europe. As can be readily seen in this book, transaction cost theory has been used as a vehicle for developing an efficiency approach. However, this approach could not be employed without first exploring the role of price in industrial marketing. Some 21 open-ended questions regarding the importance of price in the competitive strategy and pricing policies were formulated for later use. It was considered vital to make a new follow-up of the situation of the case-companies in SEA. At this point, early in 1989, the economies in the area had all recovered and there was a new boom in most industries. Much seemed to have happened to the case companies since the last trip at the end of 1985.

Except in a few instances, all the earlier key respondents had been replaced by other persons. This change also made it possible to gain access to the previously uncooperative company. However, one TNC refused access this time, which meant that three of its subsidiaries could not be visited. Since this left only one company to interview in Thailand, the last field study was concentrated on Malaysia and Singapore. Thus, this book primarily covers what happened in the case-companies during the 1980s, although some new data was collected in 1991 in connection with another study in SEA. The 1991 data concerned companies that specialize in distribution, regarding which there was very little information available earlier. Thus, the case-study base has been changed several times in the process of research, as new and relevant information has been added and information collected earlier has been reinterpreted.

Marketing, purchasing, production, and organization, as well as how these activities are related to the business environment, are thus analyzed here from a transaction-cost perspective. This provides the researcher with a broad type of theory which makes it possible to study contracting in its entirety. An organization is conceived in broad terms as involving the governance of such varied local transactions as those of customer, supplier, and government transactions. This is seen as providing a better understanding of the efficiency advantages of different modes of transacting. This widens the theoretical perspective, placing strategic behavior within a broader context. A higher level of theoretical abstraction is thus achieved.

Theoretical Fitting

The approach used in the study could be labelled a "flexibility method," in that it involves "theoretical fitting."[10] Empirical data are fitted into theories at different levels of abstraction. The objective here is to improve on "goodness of fit." The approach is both inductive and deductive. It starts in a deductive manner in that the researcher began his work taking certain theories or theoretical frameworks as a point of departure. This was pursued further in the course of research through a simultaneous, many-stage process of data collection and development of theory. The theoretical formulations which were developed were confronted with new data in the same field or in some other field of research, as well as with other theories. The ultimate purpose here was to investigate the theoretical basis of the empirically oriented case presentations, and to conceptualize these findings in theoretical terms of as general a character as possible. A basic aim of the study was to clarify the conceptual boundaries of various factors and relationships contained in the theoretical framework. In the process, several concepts were redefined. It was a question of the simultaneous clustering of theories and data at certain levels of abstraction. What this represented was not pure aggregation but a holistic process, in which parts were merged into wholes or into structures and wholes or structures were broken down into parts and merged again into other structures. This research strategy partly emerged and partly was deliberately chosen. Thus, it is a research strategy which can only be described retrospectively. Among other things, this is due to the complexity of the empirical material. This is a research approach over which the researcher maintains constant mental control, both of the theory and of the data selection, as well as of how they are adapted to each other. Data is not selected for testing a theory. Neither is theory selected for simply expressing in general terms the data obtained.

Theories are employed here in three ways. First, they constitute a raw material or an input to the study, influencing the type of data collected. Second, the data collected is considered in relationship to the development of the theoretical framework, thus representing a kind of intermediary product. Third, relevant theories appear in final form as an end product, integrated in the theoretical framework. They represent the final result of a process aimed at analytical generalization.

Sources of Evidence

Most of the empirical data was collected through interviews. This has established in an indirect way the occurrence of the behaviors in question, which were not observed directly. To obtain better insight into what actually occurred, a few factories and a considerable number of offices were inspected to observe the behavior of the interviewees and of others in their work environment. This constitutes complementary information. The same is true of company documents, chiefly annual reports, organization charts, and various informational and commercial material which were collected.

The business environment was partly researched through interviews, most of them conducted chiefly at the TNC headquarters and at certain local organizations in SEA. However, this part of the study largely involved multiple sources of evidence. Informal discussions with different persons within and outside the case companies were conducted and observations were made in an effort to answer various research questions, such as that of cultural influences on industrial marketing. To gain an understanding of the conditions under which firms operate in Southeast Asia, observations of and experience with different aspects of the local environment were also of great value. In such a study in less industrialized countries it is important not only to base findings on a variety of cases, but also to visit the area several times. Contrary to information contained in the cases themselves, however, which can readily be systematized, the special knowledge gained through such repeated visits is mainly intuitive. It is achieved through a process of abstraction which is difficult to generalize.

The Interviews

As already noted, most of the information was collected through interviewing persons at the subsidiary level. For such case studies, a questionnaire was developed based on the initial theoretical framework. It contained about 120 open-ended and nonstandardized questions. Those pertaining to purchasing and production, where the theory was well developed, were very detailed. Those applying to marketing were rather broad, the aim here being to develop a general theory; in fact these were not really questions but simply points for discussion.

All of the "questions" were not put to all of the subsidiaries. The intention was to be as consistent as possible, among other things through formulating questions in the same way in each interview. However, because of the varying nature of the questions and the varying types of companies investigated, this was not possible. No two interviews were exactly alike. Generally speaking, they were a hybrid of interview and conversation regarding specific issues.[11] In view of the holistic approach which was taken regarding the cases, as well as the managerial perspective taken in examining strategic marketing, persons in various key positions were selected at each subsidiary. The highest priority was given to the Managing Director (MD) of the company, and the second priority to those next in the hierarchy, such as sales directors, purchasing directors, and manufacturing directors. In most cases the MD was interviewed. In eight cases the interview with the person in this position was complemented by interviews with one or more in other positions. Thus, in most cases the persons from the one or two top positions were interviewed at each of the companies.

Validity and Reliability Summarized

Validity is normally defined in accordance with a deductive tradition in research, in terms of whether a measuring device such as a survey actually measures the concept which is of interest. This definition is not readily applicable to the type of case study undertaken in this project, since no concepts are operationalized or measured. In a more basic sense, however, validity concerns whether a developed theoretical framework is a relevant representation of reality. It concerns how empirical data and theory are connected. Hopefully, factors of high relevance in describing and explaining industrial marketing behavior in SEA, and closely corresponding with actual behavior there, were abstracted from the empirical data through the research process described above. Factors used to explain behavior should be genuine and not of spurious character. This allows a theoretical framework to be established which contains or is linked with momentarily stable factors and patterns which provide adequate explanations of the economic behavior studied.

This goodness of fit obtained can be considered acceptable. Construct validity was furthered by using multiple sources of evidence and having key informants review draft case study reports. A chain of evidence has been established chiefly through interviews. The internal validity can be seen as acceptable on the basis of the long and careful process of fitting theory and empirical data. As indicated above, a decisive problem was to adapt the various parts of the theoretical framework to one another. High internal validity presupposes acceptable internal consistency of the theoretical framework. Since the latter is based on many and varied cases, the external validity can be thought to be high. All of the TNCs here except one had extensive experience in many parts of the world in the business areas concerned. Therefore, the theoretical framework can be assumed to be valid for not only the TNCs and LICs that were studied in the project but also for other companies.

Reliability, just as validity, is hard to define in connection with the method employed in this project. Yin (1984, 36) declares that reliability requires the researcher to demonstrate "that the operations of a study–such as the data collection procedures–can be repeated, with the same results." This conception is not applicable here, since it is based too much on the logic for how experiments should be conducted. An objectivity of that sort cannot be achieved in this context, since such a novel study as the present one is impossible to repeat exactly. The method cannot be described in a completely objective way so as to make replication possible and, even if that could be done, the situation being researched could change on a later occasion, producing different results.

The essence of reliability here is instead that the theoretical framework established by the study reflects the true and momentarily stable factors and relations found in the research situation. The researcher was aware that biases which could come about could be seen as detrimental to reliability and did his best to minimize the influence of such temporary conditions as fatigue, anger, or sickness on the part of the researcher, and characteristics of the research situation (too warm, other persons in the room, ringing telephones, or stress) or the research instrument (e.g., wrong formulations). Interview sessions were usually very relaxed, creating favorable opportunities to follow-up answers to questions. Reliability was also improved through the careful establishment of a case-study base and its review several times by respondents.

The Cases

The results presented in the book are based on interviews at 22 subsidiaries of 13 European TNCs in SEA. The 13 organizations with their subsidiaries were introduced in Chapter One.

Since the purpose of the project was to study industrial activities in LICs, I chose a multiple-case study method. The reasons for this have been referred to above. The number of cases selected is a product of several considerations. On the one hand, there was a certain initial theoretical specification of the research area, which limited the number of cases that were relevant. On the other hand, the uncertainty surrounding the applicability of the theories in LICs called for as broad an approach as possible. Resources and opportunities to gain access limited the number of theoretically relevant cases that could finally be selected. At the start of the project, the approximate number of cases which were relevant and available could be determined on the basis of general criteria. However, the exact number later included was more a result of the research process.

The main purpose of the study was to consider as broadly as possible various industrial activities such as the marketing and manufacturing of various products within different industries and in different countries. Most of the subsidiaries were therefore to have had extensive experience of the area and were to belong to large TNCs active over large portions of the globe. At the same time, others which were smaller, less global, and newer in SEA were likewise to be studied.

As implied in the discussion above, not all the companies considered have experience from each industrial marketing field of interest. The results presented in Chapter Seven, for example, are based on interviews in 1984, 1985, and 1989 with persons in twelve subsidiaries of eight TNCs and in their group headquarters in Europe. The TNCs which specialize in distribution, two in all, are only taken up in Chapter Six, in the section on distribution and distributor specialists. Due to the considerable variation in activities, such as product breadth and depth, among the TNCs studied, the type of information collected from the different companies varies a great deal. As indicated above, not all the interview questions could be used for any given company. In addition, the answers given to the various questions vary considerably. This variety is reflected in how the case-study base is used in the book. Generally speaking, the broader the marketing activities of a TNC are, the greater the extent to which illustrations of them are presented in the book. It is essential, in this connection, that one stress the difference between such illustrations and the theoretical patterns based on the cases themselves. Such patterns have been derived through consideration of the entire case-study base, whereas the illustrations are simply drawn from particular data it contains, with the aim of clarifying certain important aspects of the theories. In accordance with the conditions stipulated in and prior to the interviews, the entire case study and its informants remain and will remain anonymous.

NOTES

1. This is a long-standing dispute in economics. Most of the debate here between different kinds of institutional economists, be they neoinstitutionalists, new-institutionalists, or just plain institutionalists, seems to concern how much the researcher should distance him/herself from neoclassical theory.

2. Williamson is not very lucid on this matter. In Williamson, 1981 (1551), for instance, one can read that organizations aim at minimizing transaction costs and that only one type of optimal organization (institution) is available. In Williamson 1985, on the other hand, transaction-cost theory is dealt with quite differently: "The crudeness of transaction cost economics shows up in at least four ways: The models are very primitive, the tradeoffs are underdeveloped, measurement problems are severe, and there is too great a degree of freedom" (390). The latter view, similar to that represented in the present study, is a much more realistic view of the conditions which apply to transaction-cost analysis. However, the question then is how much is left of Williamson's claim of having launched a neo-institutionalist economic theory. In later writings, Williamson (1986, 1989) tries to salvage the latter by replacing a causal explanation with a functionalist one. He implies, for example, that the rise of the M-form fulfills the conditions of a thorough functionalist explanation, allowing it to be given a socio-biological explanation and to be compared with the process of natural selection in the theory of evolution, or the survival of the fittest. However, this argument is not convincing, particularly since his source of inspiration, Elster (1986), doubts the validity of functionalist explanations in social sciences and instead argues for intentionalistic explanations.

3. If the attempt is made to test the assumption of perfect competition in a neoclassical model, the result is nothing more than a self-fulfilling explanation without empirical relevance, which can be used to explain practically everything. A good example of this is the use of utility theory to explain optimal foraging by animals. The decision making of animals here is explained through use of indifference curves and budget lines. The animals studied as economic consumers here, however, are not our closest relatives, the apes, but rather worms, shrimps, birds, and bumblebees (bumblebee economics). See McFarland 1985, chapter 24.

4. "By a teleological approach is meant one in which a purpose, which is not explicitly intended by anyone, is fulfilled while the process of fulfillment is presented as an inevitable sequence of events. Originally, the purpose was explicitly God's purpose unfolding itself in history. But with the growth of rationalism 'nature' replaced God, and later such entities as 'Zeitgeist,' 'history' itself, 'progress,' and more specific notions such as the 'invisible hand,' the 'market,' 'the logic of events,' appeared as secularized versions of Providence. Common to these various approaches are three features: inevitability, unintended purposiveness, and implicit valuation (though not necessarily that of the writer). The suggestion of inevitability gives the stream of historical forces a stickiness that reduces greatly the scope for maneuver, both in the past, ruling out hypothetical alternatives, and in the present, ruling out planning. The unintended purposiveness introduces terms like 'natural progress' and 'growth,' in which valuations are disguised as descriptions, teleology as causality, and reason as nature" (Myrdal 1968a, 1851).

As Hunt (1983, 101-8) indicates, teleological explanations are hard to distinguish from functionalist explanations when by function is meant "some generally recognized use or utility of a thing." For biologists the term function refers to organic processes or to vital functions such as reproduction. Anthropologists such as Malinowski define culture in a similar way, emphasizing that culture and everything it consists of fulfills vital functions. A third meaning of function is to signify "the contribution that an item makes or can make toward the maintenance of some stated characteristic or condition in a given system to which the item is assumed to belong. *Thus, functional analysis seeks to understand a behavior pattern or a sociocultural institution by determining the role it plays in keeping the given system in proper working order or maintaining it as a going concern*" (p. 102). We apply the term functionalist to such unintended purposes. Intended purposiveness is the kind of explanation sought in the present context.

5. "Economists might learn something from sociologists here. Parsonian functionalism was the dominant theoretical paradigm in sociology from World War II through the mid-1960s. A thorough statement and review of functionalism is far beyond the bounds of this introductory essay, but, to simplify somewhat, its major thrust was to explain the existence of particular institutional forms in society by reference to the functions they served for the social system as a whole. For instance, the classic functionalist explanation of religion is that it provides social solidarity. The main problem with functionalist explanation is that it is almost always possible to tell an ex-post story about why a particular institution is functional for society but far more difficult to do so ex ante. Similarly, it is difficult to connect institutional change with changes in functions. Interest in this theory has lessened in recent years as a result" (Winship and Rosen 1988, 8-9). Hunt (1983, 101-8) agrees with these views.

6. This approach invokes a number of basic problems in research. One problem is that, when the neoclassical allocation theory is dropped, efficiency can be assigned no concrete value, even if the market continues to provide an efficiency norm of sorts. How inefficient can a TNC become, for example, before internal transactions are externalized? Effectiveness and external control thus become highly complex concepts.

Another problem concerns the comparative advantages of this approach and of others concerned with explaining organizational structure and organizational growth, such as approaches concerned with power (Pfeffer 1981; Perrow 1986; Larsson 1985) and those concerned with institutional factors (Scott 1987; DiMaggio and Powell 1983). Some researchers have tried to test the validity of various theories in this area on historical data, e.g., Fligstein (1985) and Palmer et al. (1987). However, no definitive results regarding the suitability of different theories have emerged. Since no theory can objectively be seen as "best," it can be left up to the individual researcher to make his or her own choice of a theoretical approach. In the present view, to avoid tautology in research, theory selection should become part and parcel of the research process itself and not be carried out separate from empirical work. This suggests strongly the use of a case-study approach.

7. See, e.g., Glaser and Strauss (1967) for an account of such an approach to research.

8. See Bulmer (1979) and Smith, Whipp, and Willmott (1988).

9. This method differs from that of the more pure process of generalization advocated by Yin (1984), who compares the case-study method directly with the carrying out of experiments. In the multiple-case approach, he envisions a theory being replicated case by case and its external validity gradually being extended. That is mainly a deductive approach, one that demands a sophisticated theory. The premises must be stated and propositions clearly formulated so that they can be tested through experimentation. My method is neither that nor a pure abstraction process in the tradition of Glaser and Strauss (1967) or of Spradley (1979).

10. Bulmer (1979) compares this process of aligning theory and data to a zipper.

11. An excellent account of the skills needed by a case study investigator, skills applicable to the present project, is to be found in Yin (1984, 55-59).

- "A person should be able to ask good questions–and to interpret the answers.
- A person should be a good 'listener' and not be trapped by his or her own ideologies or preconceptions.
- A person should be adaptive and flexible, so that newly encountered situations can be seen as opportunities, not threats.
- A person must have a firm grasp of the issues being studied, whether this is a theoretical or policy orientation, even if in an exploratory mode. Such a grasp reduces the relevant events and information to be sought to manageable proportions.
- A person should be unbiased by preconceived notions, including those derived from theory. Thus, a person should be sensitive and responsive to contradictory evidence."

Chapter Three

The Institutional Framework

In Chapter One marketing was defined as organizing transactions to bridge gaps between industrial needs and their solution with the objective of reducing transaction costs. It was further observed that marketing transactions differ depending upon the economic institution in which they take place. The method of selling a product in a traditional market is very different from that of selling on the basis of relationships already established with a long-time customer. Therefore, a theoretical framework which applies to different types of institutions needs to be developed within which marketing behavior can be studied. Similarly, it is necessary to carefully examine different types of institutions within which internal transactions can take place. Such economic institutions create the prerequisites for the organization of vertical transactions of both external and internal form. They also involve differing ways of formulating and implementing marketing and purchasing strategies.

Regarding external transactions, it is essential to note the distinction between market behavior and marketing behavior. Market behavior is studied by economists and relates to how companies act in more traditional market forms such as monopolistic competition or oligopoly of various sorts. Marketing behavior, in contrast, is a broader term and is analyzed in this book in terms of external governance forms (Williamson 1979; 1985). It involves behavior toward both customers and competitors, irrespective of the market type. This is a vital distinction since not all industrial marketing takes place in markets, particularly not if we define markets as orthodox economists do, as a place where transactions take place and commodities are exchanged (Hodgson, 1988, 172-74). Their definitions are vague, since they take the markets themselves for granted, and need to be extended to include specification of different types of market exchange and of the institutions upon which such exchange is dependent. Consider this definition given by Hodgson (1988, 174):

> We shall here define the market as a set of social institutions in which a large number of commodity exchanges of a specific type regularly take place, and to some extent are facilitated and structured by those institutions. Exchange, as defined above, involves contractual agreement and the exchange of property rights, and the market consists in part of mechanisms to structure, organize, and legitimate these activities. Markets, in short, are organized and institutionalized exchange. Stress is placed on those market institutions which help to both regulate and establish a consensus over prices and, more generally, to communicate information regarding products, prices, quantities, potential buyers and potential sellers.
>
> Some institutions within the market are associated with exchange and contracts in an elemental sense (such as the legal system and the customs which govern the contract). . . . These would be present even if a formal market did not exist. Other institutions are specifically to do with the development of a market, and the coordination of a large number of exchanges in an organized manner.

Hodgson provides a definition of a commodity market here, but the description fits other markets as well. This is indicated below in the discussion of how institutions for economic exchange can be dealt with more adequately. The discussion takes as its starting point Williamson (1979, 1985), who defines economic institutions of capitalism as governance forms.

ECONOMIC INSTITUTIONS

The question of how transactions are organized is at the center of Williamson's analysis. A specific economic organization is considered viable and stable if its total transaction costs are minimized. Transactions are internalized in a firm when the transaction costs of using the market are perceived as being higher than those of organizing them through managerial controls. Market failures internalize transactions. Organization failures, on the other hand, in which the transaction costs within the organization are higher than those in the market, externalize transactions.

Technological dependence as a motive for vertical integration is important only in certain process industries; a more important motive in most industries is transaction costs. This is because bargaining between parties is time-consuming and expensive, making it more efficient to integrate manufacturing within one organization. Transactional dependence is thus more important than technological dependence here, and it represents the major motive for organizing technologically independent production units within a given hierarchy. Thus, human factors play a more critical role than technological factors in determining the ways in which industrial coordination is performed or organized.

There are four basic forms of organizing transactions, called governance forms or structures (Williamson 1979): market governance, bilateral governance, trilateral governance, and unified governance. The first three are external market forms, whereas the last is an internal hierarchical form. Market governance and unified governance are identical to the market and hierarchy forms, respectively (Williamson 1975). Bilateral and trilateral governance are intermediate between these two extremes. A governance structure is considered viable and stable if its total transaction costs are minimized, transaction costs being assumed to vary along three specified dimensions: uncertainty, the frequency with which transactions occur, and the size of the transaction-specific investment. These are the critical dimensions for describing transactions.

Uncertainty is an important market imperfection in the theory of governance forms. The cost of information collection and of not having full knowledge are essential costs for organizing transactions. Internal organization has certain advantages here compared with market organization. By controlling the information supplied to various persons, internal organization makes it possible to achieve convergent expectations within an organization, thereby reducing uncertainty. It also makes it possible to take advantage of bounded rationality, since organizational decision processes deal with information in an adaptive and sequential manner, an internal organization facilitating the establishment of a common language. As a consequence, the greater the uncertainty, the more often transactions occur. Similarly, the larger the transaction-specific investment becomes, the higher the transaction costs become as well.

The *asset specificity* of investments concerns whether the investments are bound to a certain type of linkage or relationship, for instance in order to realize least-cost supply (see Chapter Four). All types of investments are included, in physical and human capital as well as in production costs.

Opportunism and bounded rationality are key behavioral assumptions in Williamson's model. *Bounded rationality* is the model's assumption regarding the cognitive sphere. "Transaction cost economics acknowledges that rationality is bounded and maintains that both parts of the definition should be respected. An economizing orientation is elicited by the intended rationality part of the definition, while the study of institutions is encouraged by conceding that cognitive competence is limited" (Williamson 1985, 45). *Opportunism*, in turn, "refers to a lack of candor of honesty in transactions, to include self-interest seeking with guile" (Williamson 1975, 9). When opportunism and small-numbers conditions are linked, certain advantages in internally organized transactions as compared with market mediated transactions can be observed. In the former case, it is easier to control the opportunism of subgroups and to achieve the common goals of the organization. An organization audits the actions of the individuals in it and settles disputes between them effectively.

Just as in Alchian and Woodward (1988, 67-69), the distinction is made here between two types of opportunism: holdup and moral hazard. Holdup costs relate to resources which are tied to each other and cannot be redeployed without a loss in value. These costs are thus directly related to transaction-specific investments, as costs incurred in an effort to safeguard such investments. The occurrence or nonoccurrence of a moral hazard loss, in turn, depends on whether or not persons can be counted on to do as they promise.

The other major external factor in the model besides uncertainty is that of *small-numbers*. This concept is never really defined by Williamson. Evidently it is an expression of the number of parties to a transaction. As shown by Leblebici (1985, 101-2), this lack of specification can be traced to the fact that Williamson never explicitly defines a transaction.[1] If it is defined as he does at one point, as "an outcome of two social entities in an exchange situation" (102), the distinction can be made between an actual and a potential transaction. An actual (completed) transaction takes place between two social entities, whereas a potential transaction requires a greater number of parties. "In a market situation, for instance, we need at least two potential buyers and sellers to determine the future values of a transaction and to understand the process of exchange within the limits of opportunities and powers of each exchanging party. . . . Since each actual transaction is an outcome of a potential transaction, every transaction requires at least these four parties. Four parties are necessary to insure an efficient allocation of resources" (102). Small numbers, when the efficient allocation of resources is threatened, could be defined as the situation in which one of the potential parties is excluded from the process of transaction. Large numbers could be said to then exist in all other situations. From this it can be concluded that the small-numbers concept is an expression of how transaction costs are affected by the number of parties in the market, that is, by the horizontal market structure. The number of competitors, often operationalized in the form of concentration ratio, is a common measure of this aspect of market structure. It provides an indication of a structural imperfection in the market.

Williamson's main conclusion is that hierarchies are frequently more efficient than markets in effectuating transactions since they are better able to counteract the effects of opportunism, small-numbers exchange relationships, and information impactedness. Conflicts, therefore, tend to be more efficiently solved within an organization due its formal authority structure. Information impactedness there is less because of its better information and control systems. Organizations are superior to individuals as well in the compiling and processing of complex information, due to the sequential nature of such a process. Organizations also tend to be based on trust and cooperation, which limit opportunistic influences. The disadvantages normally associated with the market can also appear within an organization, however, particularly when the organization is exposed to a high degree of opportunism, information impactedness, and uncertainty. A small-numbers exchange relationship does not really disappear completely when a transaction is internalized. Rather, the incentives of the parties are changed. The quasi-market of a small-numbers exchange relationship within an organization increases internal transaction costs and may make it more rational to shift transactions to the market if there is a functioning market, in which other companies are prepared to conduct transactions at lower cost.

CONTROL MECHANISMS

Behavior within a governance structure is controlled by different mechanisms. For individuals to act in a coordinated manner toward a specific end, they must have information regarding suitable and unsuitable actions. Three main carriers of such information can be distinguished: price, rule, and praxis. Larsson (1989, chapter 5) employs these categories in analyzing the internal control within a large corporation. One can assume, however, that the concepts are applicable to governance forms generally.[2] *Price* was described above as the main information carrier in markets, as individual actions in markets are mainly controlled and coordinated through prices. Price is quantified information (numbers). *Rules* are a kind of basic standard for behavior, expressed verbally in text form. Rules are also formal. *Praxis*, on the other hand, is informal guidance regarding how to act. It is stored in the form of knowledge, usually of a tacit kind. It also includes activities which carry established, learned know-how. Praxis consists of implicit rules based on knowledge of how the governance form (organization) in question functions. These three control mechanisms are related to habits, which can be defined as routinized behavior. Customs is a synonymous concept. On the one hand, the control mechanisms are supposed to result in habitual behavior; on the other, habits are also transmitted to other individuals through prices, rules, and praxis. According to J. Kornai, "Stabilized and routinized behaviour establishes and reproduces a set of rules and norms 'fixed by habit, convention, tacit or legally supported acceptance or conformity'" (quoted in Hodgson 1988, 132).

EXTERNAL GOVERNANCE FORMS

The various external governance forms are discussed below and are summarized in Table 3.1. Each of them deals with the external control of vertical transactions in intermediate markets.

> The basic reason for all vertical control is the absence of a "perfect" intermediate market–a market with zero transaction costs, perfect information, and competitive pricing. To the extent that these conditions do not exist, the use of vertical control becomes more attractive relative to reliance on markets. When market costs reflect the social opportunity cost of using markets, public policy can remain neutral toward the degree of vertical control chosen. However, if markets are imperfect due to government policies or market power, there is a case for public intervention. (Warren-Boulton 1978, 3)

TABLE 3.1. Main characteristics of the external governance forms.

	MARKET GOVERNANCE	TRILATERAL GOVERNANCE	BILATERAL GOVERNANCE
Main type of information carrier	Price	Rules	Praxis
Extent of bargaining	Low	High	Moderate
Main type of enforcement (controls)	Formal	Formal/Informal	Informal/Formal
Main type of transaction	Discrete	Occasional	Repetitive
Main type of 'contract'	Rules	Agreement	Relation
Main type of trust	Organizational	Organizational/ Professional	Individual/ Organizational
Importance of trust	Unimportant	Important	Very important
No. of parties	Many	Very few	Few

The market concept is many-faceted and is in need of further specification. An important distinction is that between the vertical and the horizontal dimensions in a market. External governance forms pertain to vertical market structures. These in turn are affected by horizontal market structures and by structural market imperfections. In the model presented here horizontal markets are related to vertical transactions through the "small numbers" concept. When transactions take place between buyers and sellers operating within the framework of horizontal market structures with characteristics resembling those of perfect competition, there is market governance. This is a "large numbers" market situation. The other types of vertical control under competitive conditions are called bilateral and trilateral governance. For these, the horizontal market structures are narrow. In vertical integration or unified governance external control is replaced by hierarchical control. Decisions then are mainly controlled by fiat rather than by bargaining. These modes of vertical control of transactions are seen as contractual arrangements which are mainly governed in economic and legal ways.

Market Governance

Sometimes there is no point in differentiating between the vertical and horizontal dimensions of markets, since they coincide. The best example of this is the perfect market, where exchange is costless. However, an ideal market of this type has nothing to do with reality. Market governance as defined by Williamson (1979) is a better theoretical representation of a price-dominated market institution. It comes close to the perfect market, but the congruence is not complete, since there are transaction costs. It is a highly formalized market in which information is not sought about potential buyers and sellers, since these are lacking by definition. The assumption of atomistic competition contained in the perfect-market model is retained. The identity of individual actors in the market is thus irrelevant. Discrete transactions are emphasized. The closeness to perfect competition is evident in the fact that the entry and exit to the market is free; hence, there is no need to regulate this. Knowledge of the market is also of a high level. Even the future is predictable and can be related to the present through information about prices, products, and quantities. As can be seen above, other aspects have been excluded. Such assumptions makes it possible to specify contracts in detail, stipulating the rights and obligations of the parties under various circumstances, for example. Market governance represents the main governance structure for non-specific transactions both in occasional and in recurrent contracting. It is based on classical conceptions of contracting, in terms of which "third party participation is discouraged. The emphasis is thus on legal rules, formal documents, and self-liquidating transactions" (Williamson 1985, 69). Note that the functioning of a specific legal system is assumed, in Williamson's case, to be the U.S. system. This institution can differ from one country to another.

Market governance is thus very abstract and comes rather close to the markets found in the pure competition model. It comes too close, however, which makes the model unrealistic for the type of marketing analyzed in this book. In the present context, the condition that the future be completely known is relaxed in order to make market governance more similar to Hodgson's less formal definition of the market as presented above. Still, the most important characteristic of Williamson's definition of market governance has been retained, namely that most of the information is contained in the price. Price is the dominating control mechanism. It is a carrier of information about products and quantities, as well as about incentives, coordination, and the allocation of resources. Hence, we redefine market governance in accordance with Hodgson with the purpose of making it based more directly on real commodity markets. In commodity exchange, for example, trading is structured and information is published in order to help in the formation of price expectations and of norms. However, price is not completely determined by the market. When the degree of uncertainty is high, as in complex and volatile markets, guideline information is published so that agents can cope. There are often informal trading networks which also help to establish trading conventions and norms.

> It is because prices are stable, and are perceived by agents to be in equilibrium, that the task facing market institutions is less daunting in this respect. However, market institutions may still have many other functions, such as providing information

> regarding quality and the location of potential buyers and sellers, and regulating both the product and the entrants to the market. In fact, a crucial function may be more subtle; by ordering trade under the aegis of some institution, the price and quality of the product may be legitimized at its given level. There is a kind of stamp of institutional approval which may contribute in a powerful manner to the emergence of price norms. (Hodgson 1988, 185-86)

Market governance is signified by discrete transactions, where price is determined by the market itself and cannot be influenced by any single buyer or seller. Since commodities and raw materials are important in industrial marketing, we are interested in as precise a description as possible of this governance form.

Markets for standard industrial products such as chemicals, carbon steel, computer terminals, and shipping services resemble commodity markets in that price is the dominant information carrier. In addition, relationships are unimportant, transactions are repetitive, and switching costs are low. The "always-a-share" model described by Jackson (1985) illustrates this market type. That model is characterized by the presence of customers who purchase products from several different sellers and conditions under which the costs for shifting supplier sources are low.

> The customer purchases repeatedly from some product category. It assumes, however, that buyers can maintain less intense commitments than they do in the lost-for-good model and that they can have commitments to more than one vendor at the same time. The account can easily switch part or all of its purchases from one vendor to another, and therefore it can share its patronage, perhaps over time, among multiple vendors. (Jackson 1985, 14)

Markets for such products depart from commodity markets in one major way, however. The number of sellers and buyers is less, which means that prices are not completely determined by the market and are uninfluenced by individual market actors. Nevertheless, it is still the market which mainly governs, as illustrated by the quotation above. Although far from Williamson's original definition of market governance, which is closer to the perfect market model, such markets approximate market governance. They resemble that form much more than they do bilateral or trilateral governance.

Trilateral Governance

In market governance, dependencies between industrial units are low and relationships are unimportant. As discussed in Chapter One, however, relationships play a central role in industrial marketing and purchasing. This is discussed in connection with the trilateral and bilateral governance forms which are analyzed next. Most research based on the inter-organizational approach and a business-to-business marketing approach can be linked with these two governance forms, for example Jackson's (1985) "lost-for-good" model which dominates her book.

> The lost-for-good model assumes that a customer repeatedly makes purchases from some product category over time. At any one time, the account is committed to only one vendor. The account faces very high costs of switching vendors, and consequently it changes suppliers only very reluctantly. The account is likely, though not certain, to remain committed to its current supplier. The lost-for-good model assumes that if a customer does decide to leave a supplier, the account is lost forever–or, alternatively, that it is at least as difficult and costly for the vendor to win back such an account as it was to win the customer in the first place. (Jackson 1985, 13-14)

The more uncertainty increases, the more incomplete contracts become, as can be seen in the case of long-term contracts. Future contingencies and adaptations cannot be foreseen or considered in formulating such contracts. As compared with market governance, a more complex situation is present, there is less knowledge of the future and the actions of the individual parties matter more. Transactions are specific and occasional and are not discrete. The liberty of movement of the parties is reduced through the investment in specific assets. To increase the security of such incomplete contractual situations, the assistance of a third party may be utilized to help solve disputes and evaluate performance. This is similar to neoclassical contracting, in which arbitration is sought instead of litigation. The parties subordinate themselves to the ruling of a party outside the relationship. This third-party assistance is voluntary and is not forced upon the agents by any outside influence such as the government, although the state may set up an institution for arbitration. It is up to the parties to an agreement to decide whether or not they will use it. "The third party assistance, e.g., by arbitrators, external consultants, or advisors, can also be considered as a striving towards a situation with higher inter-party independence. The parties agree to subject themselves to external forces, which can be seen as an attempt to reintroduce a market-like situation in this respect" (Hallén 1982, 30).

Occasional transactions of mixed and highly specific kinds are then regulated by trilateral governance. The complexity of the situation, with the presence of high uncertainty and complex transactions, makes it impossible for most of the information about the product and the market to be expressed through price. On the contrary, individual buyers and sellers come to the forefront. However, the uncertainty is so great that they cannot agree completely. Assistance from outside is needed. In order for this to be successful, the parties have to work out some kind of agreement that can be accepted by the outside arbitrator. The emphasis then is instead on the individual agreement (the occasional transaction). Such third-party arrangements make it possible to formalize agreements in a contract which is then legal both according to the law of an individual country and according to international law.

Consequently, transactions take place in the development of a formal agreement here, and interactions subsequent to the agreement are guided by it. Such an agreement is specified as pertaining to a certain number of transactions which might normally be based, for example, on more general international law. Trilateral governance is an institution within which very complex products such as machines and projects are exchanged. Supplemented by rules as an additional important control mechanism, price here is one of several different conditions of exchange that are agreed upon. These prices and rules are usually determined in negotiations. An agreement becomes a major control instrument for the relationship. As seen above, uncertainty and opportunism may also be reduced by the strengthening of the relationship as such. This can be achieved through organizational trust (reputation), or through individual trust (social trust based on cultural affinity). Hence, praxis is also a part of trilateral governance.

Transactions within the trilateral governance form, such as for a custom-built piece of machinery, are not included in Hodgson's market definition, since they are considered to be non-market exchange. Trilateral governance, as Williamson defines it, is a further specification of the legal aspect of a market institution.

Both trilateral governance and bilateral governance take place in more imperfect horizontal market structures than does market governance, as in oligopolistic markets, for example. As mentioned above, a major connection between vertical and horizontal structures can be expressed through the principle of fundamental transformation, such that the establishment of vertical relations limits, or at least segments, the horizontal market–the number of buyers and sellers involved, for example.

Bilateral Governance

As with trilateral governance, transactions within the bilateral governance form are not included in Hodgson's definition of a market, since they too are considered to represent non-market exchange. A habitually renewed contract to supply a product or service to a regular client is an example of such a transaction, where already-established relationships form the basis for transactions, instead of the parties going to the market. "There is a clear distinction between regularized exchange, or 'relational contracting,' of this type, and using the institution of the market as an environment of more immediate competition and choice" (Hodgson 1988, 177). This vital distinction between relational contracting and traditional market transactions is subscribed to in the present context as well, although the conclusion drawn is different, as both bilateral and trilateral governance are viewed as exchange which occurs through the market, although not in the form of market governance. The shift from market governance to bilateral and trilateral governance can largely be expressed through the concept of a fundamental transformation which creates first-mover advantages (Williamson

1975, 1985). The horizontal market is transformed then from a large-numbers exchange situation with several competitors to a situation of small-numbers exchange with relations confined to a limited number of market actors. Also in line with Williamson, the main distinction is between external and internal transactions and not between market and non-market transactions. This approach is particularly useful in that it is possible to give a clear definition of the institutional structure of this governance form.[3]

Bilateral governance is thus completely relation-oriented. The main difference as compared with the relationships found within trilateral governance can be seen in the greater importance of the time dimension. Projects are intricate solutions to complicated needs. Linkages there are established in particular for investigating the needs involved, for preparing and implementing solutions, and for following-up. All such activities center around one lengthy and large individual business deal, such as in the case of the occasional transaction. In bilateral governance, on the other hand, linkages constitute a framework for repeated business.

Transactions characterized by market governance or unified governance are more formalized than bilaterally governed transactions. The more specialized a transaction, the greater the propensity for a hierarchical organization to be formed. Parties acting in bilateral governance structures adapt themselves more and more to each other. Recurrence of transactions over a long period of time provides the basis for mutual adjustments. As shown below it also leads to the development of routine procedures and to the establishment of trust. Bilateral governance is based on relational contracting, where the focus is on the relation between the individual parties, not the discrete transaction or the agreement. The buyer and the seller cannot regulate their business through legal rules or outside parties to the same extent as they can in the other two governance forms. The possibilities of finding alternative parties through the market decrease as they become more locked into the specific relationship. This makes them increasingly dependent upon each other. Transaction-specific social relationships are established and habits develop, for example, as a special language. Relationships built from intermediate recurrent transactions are distinguished by their low degree of formalization. The exchange is not determined by formalized rules such as in market exchange or by contract and third party assistance as in trilateral governance. Neither is it regulated through the formal rules and informal praxis of a hierarchy. The role of formal contracts in governing exchange is insignificant here, since they are impossible, or at least uneconomical, to formulate in a complete way. Price is thus chiefly supplemented by praxis as the major control mechanism.

Bilateral governance is characterized by informal rather than formal sanctions. A certain minimum of confidence and respect follow from recurrent transactions. Expectations of repetitive business restrain the parties from seeking advantages in the short run. The traits of the individual parties are essential. The reputation of a company, for instance, must be preserved. Bad will can have an impact on other business areas and increase transaction costs there. Ian Macneil believed that the relation takes on the properties of a "minisociety with a vast array of norms beyond those centered on the exchange and its immediate processes" (quoted in Williamson 1985, 71-72). Behavior within this minisociety is molded by customs or by tradition, including values. These customs and traditions are akin to laws (rules) to which the individual parties conform. Actions are routinized in that individuals acquire habits. Business is not seen as representing a process of continuous adaptation to changing conditions; rather it exists and takes place through a network of contacts which are bound by rules or routinized arrangements of a formal or informal nature. We are now far from neoclassical theory and its assumptions of non-habitual and non-routinized economic behavior, involving rational calculation and marginal adjustments in approaching an optimum. The customs and rules involved can vary from country to country and have to be learned by foreign businessmen in an unfamiliar land.

Hence, bilateral governance better controls opportunism, uncertainty, and small-numbers exchange relationships than do market governance or other governance forms under such circumstances. There are compensatory elements at play which institutionalize transactions and reduce the occurrence of actions contrary to the relationship. Such soft contracting, to be viable, "needs to be supported by a more elaborate informal governance apparatus than is associated with hard contracting" (Williamson and Ouchi 1981, 1538).

TRUST

A habitual behavior does not occur if individuals do not trust each other or trust that the rules of an organization will work in the manner expected. The distinction will be made here between organizational trust and individual trust (see Figure 3.1). Organizational trust originates in the organization, for example, in trusting its rules to work in the expected way, or in the buyer trusting the seller as an organization to live up to the promises made. This can be expressed as *reputation*. Individual trust has to do with persons and the friendship between them. This trust is of two kinds. One concerns the individual as a representative of his or her company when friendship originates in conjunction with the completion of working tasks. It then is professionally determined and is a consequence of the necessity to meet in order to solve common business problems. This professional trust is more instrumental than emotional in character. Another kind of trust concerns bases of friendship other than business, such as personal traits, membership in a social or cultural group, or the like. This is called social trust: persons associate because they like each other, or because they belong to the same social or cultural group. The relation between social trust and emotions is complex. Where trust is based on personal traits, friendship is definitely based mostly on emotions. However, where social trust comes from a common social or cultural background, a low degree of emotional affiliation may be involved. As will be seen below in discussing the impact of Chinese culture on business relations, for example, "guanxi" manifests itself as a mixture of instrumental and emotional (expressive) components. Social trust is used here in an instrumental way, to further the ends of business.

The importance of trust varies with the state of the relationship. In the beginning it tends to be slight and can mainly be classified as organizational. This varies, however, with the situation. As shown in later chapters, in SEA it is frequent that no such trust exists. Even very large TNCs lack reputation, since they are unknown in the area. Therefore, organizational trust has to be built from the ground up, usually as individual trust that later develops into organizational trust. One comes to know and trust the party with which one deals, through contacts with individuals working with the organization in

FIGURE 3.1. Different kinds of trust.

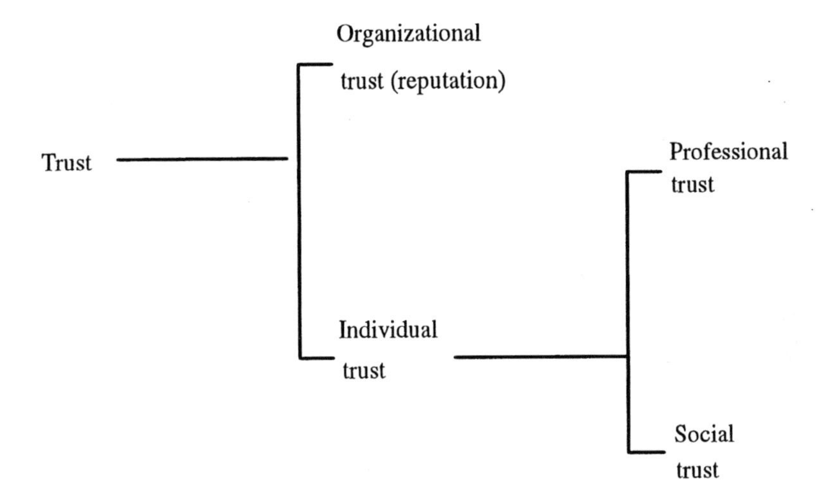

question. Trust has to be established in order for behavior to become routinized.

Consequently, there is both a personal and an organizational stake in linkages characterized as bilateral governance which mitigate the opportunistic inclinations found in the rest of an organization. Parties directly involved in a specific relationship can prevent others in the organization from unduly misusing the relationship. The more dependent the parties are upon each other, the more strain can probably be endured by the relationship. Habits strengthened by trust are costly to change.

As will be discussed in more detail in Chapter Four, the cost of dependence is expressed through the concept of linkage specificity, of which social impactedness is one dimension. Trust is primarily related to that aspect. The higher the linkage specificity, the more the liberty and movements of the firm are reduced. The more trust is able to reduce transaction costs in existing linkages, the higher the costs become for switching to other linkages. Trust is a kind of strategic compensating element which lessens the influence of opportunism. It disciplines the actors, thereby constraining behavior. Enforcement and bargaining costs are reduced. Trust also reduces search costs and thus information costs as well. The need for the buyer, for instance, to engage in search regarding the seller's offer and intentions is less if the trust in the seller is high. As a relationship develops, this attitude tends to become more positive. A positive attitude facilitates business, whereas a negative attitude has the opposite effect. Empirical studies show that trust is a dominant attitude in many industrial market relations.

Trustful relationships are perhaps particularly common within firms in which persons are not only rewarded for the market values of what they exchange, but also for their contribution to the intra-firm group to which they belong. A hierarchy may be highly effective in distributing the high fixed costs involved in establishing trust over an extended period. Market contacts are often more short-lived and afford fewer opportunities to establish personal bonds. Firms can also increase confidence and homogeneity through recruiting employees of similar backgrounds in education, other socioeconomic factors, culture, etc.

Although to a lesser degree, these hierarchical characteristics are also valid for certain external transactions, particularly bilateral governance. In bilateral relationships, the high fixed cost of transaction-specific investments involved in trust-building activities is normally spread over a long period of time. The costs are also shared among the parties. Moreover, the forming of trustful external market exchange is facilitated if purchasers and marketers in separate firms share a common background regarding culture and socioeconomic factors such as education. Social trust and professional trust work together to enhance the social impactedness of a relationship.

ADAPTATION

When environmental or business uncertainty increases in bilaterally governed relationships, changes are quite conceivable. The relationship can be deepened and adapted to an uncertain environment. Alternatively, transactions can be shifted to a market governance form by standardizing the relationship, for instance for a special product, or by integrating it into a hierarchy. If uncertainty decreases, on the other hand, changes in the opposite direction could occur.

There is the danger of inefficient adaptation to changes in a relationship of the type described, since there is always the risk of opportunistic and expensive haggling between the parties. Even though they share a common long-run objective regarding the project, each party also has an interest in getting as much as possible in the short-run (Williamson 1979, 251).

One important question is how pervasive is the trust in a relationship. Is it possible to disregard opportunism and the fact that there are no important transaction costs involved in the relationship? If trading risks are large, for example, uncertainty and a small-numbers exchange relation readily arises. A better solution then could be to internalize transactions. As discussed above, common organizational goals facilitate the harmonization of the relationship, and there are also better possibilities of sanctioning the behavior if it occurs within an organization. A hierarchy can also create an atmosphere of trust between employees more economically than a market can. The governance form selected can thus depend upon the possibilities of the respective parties for controlling the behavior of the individuals involved.

IMPACT OF NONECONOMIC INSTITUTIONS

The major characteristics of the three economic institutions are summarized in Table 3.1. The impact on these governance forms from the noneconomic institutions of greatest importance, namely government and culture, will be discussed next.

Government

The State as an institution is excluded from neoinstitutionalist transaction cost theory. However, since governments often take a direct part in organizing transactions in developing countries, they are included here. Intervention in the functioning of economic institutions can take two main forms. First, government regulates the premises which face economic institutions by manipulating various societal institutions, such as improving the infrastructure or the legal system. Second, government takes an active part in the governing of economic transactions, for instance governing transactions by means of a planning system or replacing private governance by state governance. In the latter case, the main question is whether this is actually more efficient or is primarily done for ideological reasons. The question then is whether government is more efficient than markets or firms.[4]

Situations in which government regulates transactions between buyers and sellers are similar to trilateral governance. They differ, however, in that the transactions are forced upon the parties and not first entered into voluntarily and then regulated by legal institutions. The control which is exercised can be direct, as when government takes an active part in controlling individual deals. It can also be indirect, as when government establishes rules to be followed by the parties to an agreement.

Culture

Cultural factors have a strong influence on governance forms in SEA. This became very evident in this project as we gained insight into the transactions that took place between industrial firms in the region. These transactions were clearly reflective of the Chinese way of doing business, something which very much dominates business in the region. In this section the extent to which Chinese business behavior can be classified in terms of the three external governance forms already mentioned, or seen as a mixture of two or more of these, will be examined. First, one can note that there are some similarities with certain of the characteristics of market governance. For instance, Chinese business deals can usually be classified as discrete and short-term, particularly since they often are made between small family firms. Price is also the main signal to the market, implying that price contains most of the information about products. However, how prices are fixed is more reminiscent of other governance forms. Bargaining is a dominant characteristic, causing high bargaining costs in transactions. Uncertainty is rather high. A similarity with bilateral governance can be seen in that great importance is attached to social relations and trust, and these relationships are highly informal. Although such relationships are strong, business tends to be done in the short term. Since social relationships are of such importance, it takes considerable time to make an individual deal. These cultural aspects also affect the long-term character of business. In Europe, in contrast, industrial marketing tends to take place within continuous and long-term relationships, with separate deals made within the framework of the linkages involved. In SEA, however, the converse is more often true. Separate deals are made and if over a number of years a whole series of such deals between the parties are concluded, this is viewed as a long-term relationship. Strong linkages can thus be described here as a long succession of separate deals. The social part of a linkage can generally be characterized as mixed rather than instrumental.

Chinese business culture, as it is expressed through such governance forms, is thus inconsistent with Western business culture. The inconsistencies derive from three sources: the long tradition of the Chinese as traders in commodities, the traditional Chinese way of doing business in bazaars, and the Chinese culture as such, which is very much based on relations. The major results of the Chinese trading tradition in the region are the individual deal orientation, the importance attached to cash, and the proneness to bargaining. Bargaining is also characteristic of bazaar trading. Such business transactions are characterized by a high information asymmetry among trading parties. Mistrust is also widespread, which results in high enforcement costs. There is no third-party enforcement. Cultural influences to some extent discourage trilateral governance. Social contact nets are founded to alleviate both the mistrust and the information gap concerning prices. For any given party it is very much a question of having a better contact net than the other party; social information is very important.

The Relation-Oriented Culture

These relation-orientational qualities of business transactions are derived from basic characteristics of Chinese culture generally, in particular the importance of the family and of other relationships, as well as a tendency toward shortsightedness. *Face* is important too, as harmony in relations is emphasized and conflicts are seen as socially unacceptable. Face is a critical part of relations in the primarily group-oriented Chinese society. It is supported by a vertical group structure that is centralized and authoritarian. The impact of Chinese culture on both the external and internal relationships of the TNCs studied was found to be of crucial importance (Jansson 1987).[5]

The relational orientation of Chinese culture dates back to Confucianism, certain of the essential aspects of which Bond (1986, 216) describes as follows:

- Man exists through, and is defined by, his relationships to others.
- Relationships are structured hierarchically.
- Social order is ensured through each party's honouring the requirements in the relationship. (Bond 1986, 216)

On the basis of their expressive and instrumental components, Bond (1986, 223) classifies interpersonal relationships here into three main categories:

1. Expressive ties: the family (or the clan) is the most important social unit in Chinese society and so expressive ties are most often found there.
2. Instrumental ties: these are temporary and anonymous or highly impersonal relationships with others in order to attain personal goals, e.g., relationships between customers and sellers in stores, banks, etc. Expressive elements here are insignificant. In such short-term exchanges (transactions) there is little chance for friendship; the trust involved is mainly of a professional character.
3. Mixed ties: "The mixed tie is a kind of interpersonal relationship in which an individual is most likely to play the power games surrounding guanxi. In Chinese culture, persons who are said to have guanxi usually share one or more important characteristics. They may be ascriptive, for example, common birthplace, lineage, or surname, or may involve shared experience, such as attending the same school, working together, or belonging to the same organization (Jacobs 1979). The guanxi outside an individual's immediate family are conceptualized as mixed ties in Hwang's framework. Generally speaking, both sides to a mixed tie know each other and maintain a certain expressive component in their relation-

> ship, but it is not as strong as that in the expressive tie." (Bond 1986, 224)

Guanxi is characterized by connections and secure personal ties outside the family or clan. Friendship is not professional but is social and yet has another connotation than in the West. As indicated above, this culturally determined friendship is, for instance, more ascriptive (quanxi) than achieved. It is neither friendship nor moral obligation.

> It is the quality of a relationship that is premised to a substantial degree on common interests. . . . The guiding principles of guanxi lie not so much in the psychological realm as in *objective factors and conventions of behaviour* [italics added]. . . . Among Chinese it is expected that people who share a common background will instinctively be mutually supportive: people who are from the same place–village, province, or region–or who attended the same school, or, better yet, were classmates, or who served in the same organizations are expected to be available to one another. (Pye 1985, 293)

According to Bond (1986, 224-225) such a shared particularism can be established in China by strengthening affection (ganqing) and pulling or working connections (la quanxi or gao guanxi). The first and most popular way is to increase social interaction by visiting, giving gifts or inviting persons to important family events and festivals. Second, a person who wants to be introduced to or solicit a favor from another person may use a third person of high social status as an intermediary, thereby opening up a "back door" to that person. A third common strategy to enhance one's influence over others is face-work. "Since an individual's power and status as perceived by others can guarantee an allocator's help, it is important for a Chinese to maintain his or her face and to do face-work in front of others in the mixed tie" (Bond 1986, 225).

Other Essential Cultural Influences

The Chinese are also known for being entrepreneurial and hardworking. These cultural traits are closely interrelated and can be referred to as Confucian values. Among other things, they are manifested in business as a low uncertainty avoidance and as shortsightedness. The same is true of two other crucial characteristics that influence business relations: pragmatism and materialism. A basic cultural factor behind these various attitudes, beliefs, and characteristics is that of causality, which can be defined as a form of explanation of the connections between events or phenomena.[6] Probability is the extension of this into the area of prediction[7] (Redding 1980). Another factor normally associated with Chinese behavior is superstition.[8] It too is related to causality but its chief relation is to probability. Each of these cultural traits has an important impact on the activities of the Southeast Asian TNCs which were studied (Jansson 1987).

CONCLUSIONS

The broad regulators of external transactions–economic and noneconomic institutions–have been presented in this chapter. Various qualities of transactions determine how they are organized. The asset-specificity and frequency of transactions, and the occurrence of opportunism and of bounded rationality, for example, result in costs for information, bargaining, and enforcement. These costs can be reduced in different ways, depending upon how decisions are made and implemented in transactions, how conflicts of interest are solved, and how decisions and conflicts are controlled. If these costs are lowered through making use of an authority mode, the decisions and the conflict resolution tend to be internalized.

Alternatively, if a bilateral or trilateral mode is of lesser cost, decisions as well as conflict resolution tend to be externalized (some of them at least). Adaptation takes place between and within governance forms. The main characteristics of external governance forms are summarized in Table 3.1. The main control mechanism in bilateral governance is praxis, in trilateral governance it is rules (formal agreement), and in market governance it is price. Bilateral governance stands out from the other forms because it is primarily regulated by an informal contract, of which trust is a major ingredient. Trust was divided into organizational and individual trust, with the latter type further subdivided into professional and social trust.

The cultural influence on social trust expressed by the Chinese concept of "guanxi" was emphasized. It was indicated that in both trilateral and bilateral governance transactions are characterized by the important role of linkages.

These economic institutions of capitalism are constrained in turn by noneconomic institutions. They are connected by means of a transaction-cost logic. Government policies, a weaker legal system, and differing ethical standards in business, for example, may increase uncertainty and make opportunistic behavior more common. Transaction costs then become higher, resulting in a stronger tendency toward internal governance. Government is an important factor to consider here, since in LICs government often takes a direct part in organizing transactions. Public fiat, for instance, is often employed in conjunction with a planning system. The state also has an indirect influence on economic institutions through the manipulation of other, noneconomic institutions, such as the legal system.

The other major noneconomic institution is culture. The relation-oriented Chinese culture affects the different institutional forms in which business activities are organized. "Guanxi" is one expression of this. The Chinese can be characterized as family-oriented, hard-working, entrepreneurial, pragmatic, clannish, superstitious and materialistic. They also have a traditional acquiescence to and desire for hierarchical authority and an incapacity to stay united in a common cause, except within their inner circles.

NOTES

1. Leblebice's article is based on the writings of Commons.

2. Larsson (1989) gives examples of control mechanisms in other institutions, e.g., personality among friends, genes within the clan, and ideology within political movements.

3. Hopefully, the distinction between market and non-market transactions will not continue to "elude most economists" (Hodgson 1988, 177).

4. Chenery (1979, 209-13) discusses three types of coordination mechanisms: integration under private control, the Lange-Lerner system of centrally administered prices, and direct control of investment. Development theory emphasizes the vertical interdependence of investment decisions. Coordination of investment projects in developing economies is vital, since a group of investments which would be profitable if carried out collaboratively may, if undertaken separately, be

unprofitable and thus not be undertaken at all by the individual investors. In order to reap the full potential of external economies it is often necessary, according to a recommendation commonly made in this context, that the government administratively coordinate such investment projects due to the inefficient or insufficient coordination of them exercised by the market. Markets here either do not function at all or function very badly. In developing countries, markets are at different stages of development, ranging from not being markets at all to being underdeveloped as compared with those in the industrialized countries of the West. The cost of running the economic system in such countries tends to be high. A bias found in a non-institutional approach to this problem is that, according to such an approach, markets are implicitly understood as involving the type of perfect competition assumed in neoclassical economics. This makes the markets of developing countries appear quite imperfect. A more realistic view takes market imperfections as a starting point and looks for the existence of various imperfect market forms such as those of bilateral and trilateral governance.

The emphasis in the industrial planning of developing countries is usually on sectors (aggregated products), production efficiency, and the basic advantages of coordination. Not much is usually said about how to organize the desired coordination, particularly on the company level. Chenery (1979, 209-13) emphasizes the importance of this question but offers only "a few scattered comments" (210).

Since most of the industrialization in such geographical areas is based on the activity of TNCs, the major focus must be on them, not on the industrial sector as such. This provides a fresh approach to industrial planning there. More usual analyses of industrial structures at different planning levels in such countries tend to focus on production costs and production efficiency. Although these are essential to consider, this must be supplemented by consideration of the costs of coordinating production units or sectors. TNC theories are thus important. How they are related to institutional economic theory in a broader sense is the focus of this book. I will suggest, for instance, that TNCs in their investments in developing countries mainly reproduce an industrial structure which has been formed in their previous operations within developed countries.

This aspect of industrial planning is discussed at greater length in Jansson (1984b).

5. This result is in no way unique or surprising. Several studies show how Chinese culture manifests itself in Southeast Asian business life, e.g., Villacorta (1976), Redding (1980, 1982), Redding and Ng (1983), Redding and Pugh (1985), Redding and Wong (1986), Hofstede (1980, 1984), Hofstede and Bond (1987), Lim and Gosling (1983), Limlingan (1986), Lasserre (1988), and Wimalasiri (1988).

6. Redding (1980, 132) describes this cultural theme as follows: "A parallel view of such a theme is taken by Nakamura (1964) [Nakamura, 1964. *Ways of Thinking of Eastern Peoples*] in one of the leading works on Oriental thinking. In this, he singles out the following characteristics as typical of Chinese thinking:

1. Emphasis on the perception of the concrete
2. Non-development of abstract thought

3. Emphasis on the particular, rather than universals
4. Practicality as a central focus
5. Concern for reconciliation, harmony, balance.

This is contrasted with Western cognition, which is described in terms of "logical, sequencial connections. Use of abstract notions of reality which represent universals. Emphasis on cause."

7. In Western culture the future can to some degree be calculated, whereas in Chinese culture there is more of a fatalistic view of the future (Redding 1980, 133).

8. Superstition is a doubtful concept in this context, since it is too much colored by a Western outlook on Chinese culture. Readers further interested in this subject are referred to Eitel (1984) and to Lip (1989).

Chapter Four

Industrial Marketing Strategies

INDUSTRIAL MARKETING AND TRANSACTION COSTS

Use of efficient marketing strategies to achieve long-range competitive goals involves a reduction in transaction costs. When transaction costs–or costs of information, bargaining, and enforcement–are high, as they frequently are in the far from perfect real economic world, price tends not to be an accurate reflection of the value of the resource exchange which occurs, since it does not express with any precision information regarding the product and the demand for and supply of it. Due both to uncertainty and to bounded rationality, buyers and sellers hardly can have full knowledge of the product and market. How much information about a product is contained in its price depends in part on the type of product. The price is a closer expression of the value of the resources exchanged in the case of technologically simple standard products. The market situation of these products can be largely characterized as one of market governance. This means that price provides relatively accurate information about the economic value of the product. Price competition is the dominant force in the marketing of such products, and tends to result in a rather efficient use of resources. The situation is different, however, for heterogeneous products such as the special, technologically complicated products considered in this book, the market conditions for which are far more imperfect. The more the economic exchange situation departs from the ideal of a perfect market, the greater are the resources which are spent in collecting information, in bargaining, and in policing agreements, and the less information-value price will have regarding the product and the market situation.

Under such conditions in particular, not all information about the product and its market situation is expressed in its price. Additional information about the product, about other products and their prices, about the availability of the product, and about potential buyers and sellers is needed in order to be able to determine the appropriateness of a particular price in a specific market situation. The individual parties involved also play an important role. In many complex marketing situations, close personal relations may develop between the parties and lead to a reduction in transaction costs. The exchange of goods here is not simply a technical matter between parties who are unknown to each other. Rather, resources have been invested to obtain and provide information about the needs involved and the degree to which the product can satisfy them. Resources can be invested in advertising, personal selling, and public relations. Information through such non-price means reduces uncertainty. Trust is another common way in which the uncertainty of the market actors may be decreased. It can be seen both as a substitute for the collection of more information and as a necessity when no further information can be obtained and a deal is to be made. One effect of the investment of resources in such trust-building activities is reputation, by which the company acquires a crucial asset: good will.

By spending resources on trust-building activities, a company tends to reduce its bargaining and enforcement costs. Bargaining costs come about in imperfect markets due to certain needs which are not easily satisfied. Resources are spent in negotiating deals, the major purpose of which is to discover or determine prices. At the same time, agreements thus made may not be followed and implemented in the manner originally conceived. Costs may then be incurred in exercising controls concerning agreements regarding prices and other conditions. The three main types of transaction costs which were mentioned, those of information, bargaining, and enforcement, are related partly to uncertainty and partly to opportunism. Opportunism of the parties clearly increases transaction costs. Bargaining costs arise, for example, if there is haggling. Costs for advertising, personal selling, and the building of trust represent both bargaining costs and information costs. They are also partly related to enforcement costs. Through personal selling, for example, customers are informed about how a product will fulfill a particular need. This increases information costs but may accordingly reduce enforcement costs. Service and guarantees are means for increasing the security of the customer and for likewise reducing enforcement costs, which are to a large extent connected with the implementation of marketing deals and purchasing decisions.

Resources are also devoted to logistics in order to coordinate production units and the transfer of goods. Skillful coordination here reduces transaction costs connected with transportation, for example freight costs and capital costs, or certain production costs. Transaction costs related to logistics can also be viewed as reflecting the accuracy of certain transactions, those concerning, for example, times of delivery and delivery security.

Other parties in the market also influence transaction costs when such questions as the following are considered: What other potential customers have a need for similar products? What other potential sellers can offer similar solutions? There is also a need for information regarding the preconditions of the participants in a transaction. The distribution of resources between competing parties can be important here.

THE TRANSACTION COST MODEL

The transaction cost model is depicted in Figure 4.1. The various industrial marketing strategies analyzed in this and the coming chapters are defined and limited by the general (generic) vertical market structures discussed in the previous chapter. Strategies are also strongly influenced by various factors which are discussed below.

Reduction of Transaction Costs

As already indicated, an efficient reduction of transaction costs is often achieved by establishing buyer/seller linkages. Bilateral governance and trilateral governance differ in the constraints they place on these cost-reducing activities.

FIGURE 4.1. The transaction cost model.

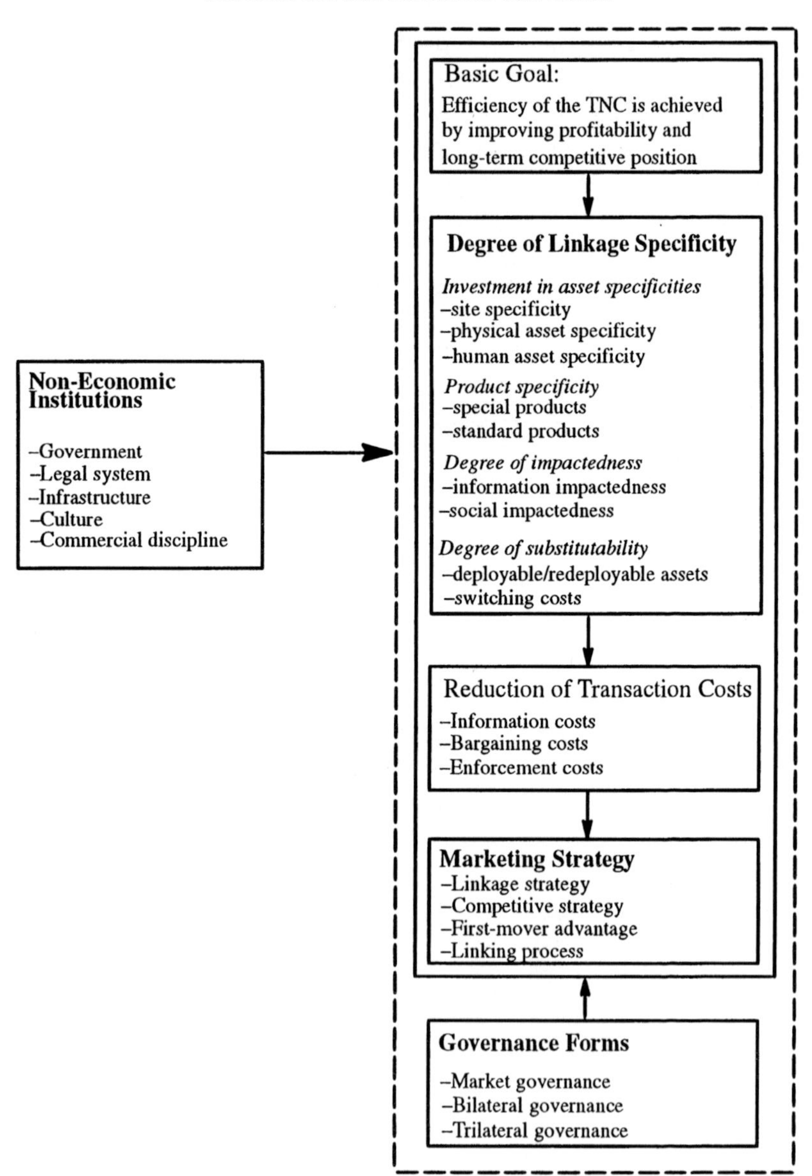

Basic Goal

The basic goal of a TNC is efficiency. Long-term profitability and viability are achieved by being more efficient than competitors.

Degree of Linkage Specificity

The focus of the transaction cost model as presented here is linkages or relationships. The investment in linkages creates dependencies or bonds between the parties. Such dependencies are given an economic interpretation here. Investment in linkages is directed at a specific linkage or at linkages in general. The former case will be referred to here as a "linkage-specific investment" which is preferred to the similar conception "transaction-specific investment."

Investment in Asset Specificities

This concerns the investment of resources in manufacturing the product or products sold to a particular customer or to a group of customers, as well as in the parts of the sales and distribution organization utilized in dealing with a specific customer or group of customers. This comes close to the asset-specificity dimension of market dedication, since it covers investments in dedicated assets on behalf of a particular buyer or a particular group of buyers. Such a linkage-specific investment could consist of investments in human assets (human asset specificity), in equipment (physical asset specificity), or in immobile assets bounded to a certain geographic area (site specificity). According to Williamson (1985, 95-96), all types of investments are included in investments in asset specificities, investments in physical and human capital, as well as production costs. Williamson distinguishes between four main types of asset specificity: site specificity–e.g., successive stages that are located in close proximity to one another so as to economize on inventory and transportation expenses; physical asset specificity–e.g., specialized dies that are required to produce a component; human asset specificity that arises in a learning-by-doing fashion; and dedicated assets, which represent a discrete investment in generalized (as contrasted with special purpose) production capacity that would not be made but for the prospect of selling a significant amount of product to a specific customer.

Degree of Impactedness

The asset-specificity principle is also related to the transformation of transactions from large to small numbers bidding situations (Williamson 1981, 1548).[1] The advantage thus achieved is referred to as a first-mover advantage, as depicted below. It can be said to be achieved through the creation of impactedness of different types. The concept of information impactedness is extended here to include the social dimension of a linkage. Social impactedness is created through social interaction. The impactedness concept is closely related to linkage-specificity: the higher the social impactedness of a relation, the higher the linkage specificity.

The Product Dimension of Linkage Specificity

This concerns to what degree a product is specific or general. Specific products are sold through specific linkages to specific customers. The opposite is the case for general products.

Degree of Substitutability

The assets involved in a specific linkage are economically bound by the linkage, being expensive to redeploy. The other party usually makes corresponding linkage-specific investments as well. Mutual adaptations take place as investment in the linkage progresses. The costs of taking advantage of these investments are much lower than they would be for alternative users. The mobility of these assets in relation to alternative linkages is low, which results in a low degree of substitutability of the linkages and of the companies connected by the linkages. The costs of transforming such assets are designated as switching costs. Thus, the higher the linkage specificity, the higher the switching costs. Furthermore, the more imperfect the market, the greater the extent to which the continuity of the relationship is valued by the parties. Contractual and organizational safeguards are created to support continued transactions within such linkages. In contrast with the mutuality in the inter-organizational approach, which is socially determined, mutuality here is viewed as economically determined. The binding of assets in a linkage is thus seen as the result of how firms react to various information, bargaining, and enforcement costs and how they seek to reduce these costs.

Governance Forms

The immediate context of this vertical marketing model, or the broad constraints for how to act and organize actions in accordance with it, are expressed as governance forms. Market governance, trilateral governance, and bilateral governance provide differing conditions for marketing strategies. There must be a congruence between the characteristics of transactions and the type of governance form involved. These preconditions can also be influenced by the marketing strategy employed. In this way, structure and strategy evolve over time in a complex pattern of mutual interaction. Companies use a particular structure or governance form to achieve competitive advantage. This calls for an intimate knowledge of the governance forms in question. Firms need to know how they can restrict certain actions of the other party or parties and how the existence of a particular governance form can be influenced.

THE MARKETING STRATEGY

In this section we discuss how firms act to create and maintain relationships involving linkage specificities. A given marketing strategy concerns how a company, here a TNC, acts toward customers and competitors in a market. This is determined by both the capabilities of the company and its aims. The advantages a TNC can achieve through use of various kinds of assets are discussed later in the chapter and a number of differing strategic marketing profiles are described. These profiles concern the capability of a company to handle different forms of linkages.

Institutions are seen as constituting the framework for different marketing strategies. The various external governance forms which have been discussed are relevant to the strategies to be covered. It can be observed generally that trilateral and bilateral governance constitute the framework for direct contacts between buyers and sellers in the types of cases considered here. Governance forms represent broad categories of conditions under which transactions are organized. They are major transaction cost patterns within which companies act. The general conditions for the organization of transactions (infrastructures) represent constraints which can be interpreted both as opportunities and as potential threats.

A marketing strategy can be said to consist of two basic substrategies: a linkage strategy and a competitive strategy (see Figure 4.2). The *linkage strategy* is related to vertical transactions and concerns how the seller creates connections with the buyer. As discussed in Chapter One, information costs, bargaining costs, and

FIGURE 4.2. Industrial marketing strategies.

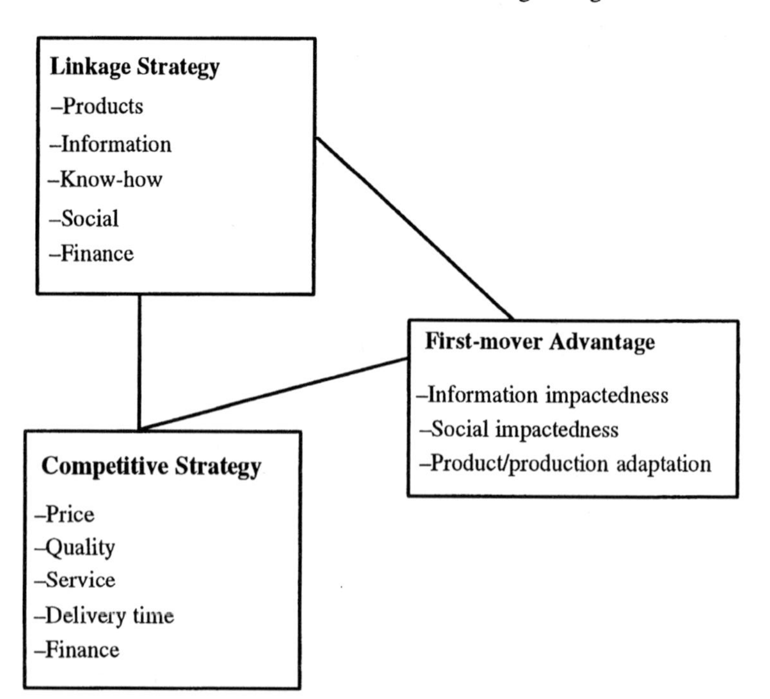

enforcement costs are the major transaction costs involved in bridging the gap between an industrial need and its solution. A marketing process allows this gap to be bridged in connection with or through the establishment of an agreement or relationship. Through a linkage strategy, the seller competes with the buyer for resources involved in the exchange. A *competitive strategy* concerns the position the seller takes or possesses in relation to competitors and whether the offer is competitive. This strategy is closely related to horizontal market structures and can be called "horizontal competition." A crucial aspect of strategy is to achieve a *first-mover advantage*, something which is based on and maintained by an efficient mix and sequence of linkages over time. The market condition is transformed in this way from a large-number to a small-number exchange situation. The number of genuine competitors is gradually reduced. From a strategic point of view, this basic or fundamental transformation process can be described as the spinning of a web so that the buyer is enclosed in the web, or relationship, at the same time that competitors are locked out.

The Strategy Concept

Strategy can be defined as "a pattern in a stream of actions" (Mintzberg and Waters 1985, 257). An advantage of this definition is that it allows a strategy to be viewed in a nonprescriptive way. It is conceived in terms of how companies *actually* decide and act, not how they *should* decide and act. What is to be considered is not how companies should be managed but how they are managed. Implications of the findings for the strategies that companies best should employ are considered in Chapter Eight. The strategic behavior of firms is considered in this book from an economic perspective and in terms of transaction cost theory developed to explain such behavior.

This definition of strategy is broader than one limited to viewing strategy as a pattern of decisions. This makes it possible to distinguish between deliberate strategies that are performed as intended, and emergent strategies that are patterns of action carried out despite, or in the absence of, intentions.

> It is difficult to imagine action in the total absence of intention–in some pocket of the organization if not from the leader-

> ship itself–such that we would expect the purely emergent strategy to be as rare as the purely deliberate one. But again, our research suggests that some patterns come rather close, as when an environment directly imposes a pattern of action on an organization. (Mintzberg and Waters 1985, 258)

No pure emergent strategy and no perfectly deliberate strategy was observed in the reported empirical work. A strategy of the latter type presupposes an environment that is perfectly predictable and is under the full control of the organization. Most strategies are intermediate between these pure forms. Mintzberg and Waters have presented a continuum of strategies. A planned strategy is in their terms the most deliberate one, whereas an imposed strategy is the least deliberate but also the most emergent. Between these two extremes there are six other strategies they describe.

Porter (1980, 1981, 1986) interprets strategic behavior from an economic perspective. He refers to competitive generic strategies conducted within the framework of the market structure. His conception is based on a so-called structure-conduct-performance paradigm. Strengths and weaknesses of firms are interpreted in terms of structural factors such as mobility barriers, number of competitors and degree of rivalry between them, the relative power of suppliers and buyers, and the potential of substitute products which influence strategy. The firm's purpose is considered to be to manipulate these factors within the market structure of an industry, either through modifying or activating them, in order to earn higher than normal economic returns on investment. The proper formulation of strategy is seen as being based on a knowledge of what structural elements to change.

Although strategy is not viewed normatively in this book, the approach taken is similar. Whereas Porter emphasizes the interdependence between market structure and strategy, the interdependence between institutions and strategy is emphasized here. Although Porter builds on industrial organization theory, the strategic approach in this book is based on institutional economic theory. Strategies are seen as being formulated and implemented here within the framework of institutions. These are divided into two main groups: economic and noneconomic. The economic institutions consist mainly of various governance forms. These are seen as constituting a vertical structure within which industrial marketing takes place. Governance forms can be considered parallel to market structure as described in industrial organization theory. The noneconomic institutions include the infrastructure, the political system, culture and social structures, and the like (see Figure 1.3).

Institutional factors are considered to determine the rules of the game and constrain industrial marketing strategies. Like the market structures in Porter's model, they can be affected by how customers and products are selected. However, the focus in this book is on vertical relations of the firm in its interface both with customers and with suppliers. Vertical rather than horizontal competition is concentrated on. Like Porter, the concept of mobility barriers is taken here as a starting point, for instance in the form of specialized assets and economies of scale. In addition, I venture beyond the neoclassical conditions of supply and demand to investigate the effect of supplier and buyer relations on strategy.

The model presented here is broader in another way. To Porter's industrial organizational competition is added Chamberlainian competition (Barney and Ouchi 1986, 372-80). One assumption of the monopolistic competition the latter involves is that firms are heterogeneous and have unique assets which allow them to compete. It follows from that assumption that strategies should exploit not only external market factors but also the unique internal skills, resources, and distinctive competencies of firms.

The focus in works based on structural market imperfections (e.g., Porter 1980, 1981, 1986) is on strategy design. Unlike Porter, a normative approach to strategic design is not adopted here. Rather, a descriptive approach to strategy and to the organization of strategies is taken. When the organization of strategies is emphasized, strategic implementation and action assume particular importance. This approach leads into the area of strategic management. Most conceptions of strategic management involve a combining of the design and carrying out of strategy, emphasizing joint efforts to formulate and implement or organize strategies, for example through Strategic Business Units. Although it can, in fact, be said that a transaction cost approach focuses directly on both of these aspects, it is not conceived here as representing strategic management because of the absence of the normative aspect.

The Linkage Strategy

The linkage strategy and the competitive strategy are the two main marketing strategies. The first of these pertains to vertical competition in particular, and is a relevant strategy when the structure of transactions involves bilateral and trilateral governance. A combination of four major types of transactions is involved, those for establishing information, social, product, and financial linkages.

Through the linkage strategy, a bond is established between the buyer and the seller. Under conditions of trilateral governance, for instance when developing a project, buyers and sellers often work together closely for an extended period to create an agreement. Through this process they become enveloped in their relationship, making the substitution of either of them by another party difficult. The seller, for instance, becomes involved in variety of the customer's concrete problems, resulting in a very specific linkage. The main way to protect the continuity of such a linkage is through an agreement which is safeguarded by a third party, such as the host or home government. Normally, however, this is not enough. The distance to be covered between the start of commitment and the results in the form of an agreement is often very long. Commitment grows gradually as the parties become more and more absorbed by the relationship. The strong mutual interdependence which results involves considerable dangers for both parties. For one of the TNCs studied, Telecom, a large order for a system means a "marriage" with one customer for at least ten years. Such a strong tie and the long-term character of large projects makes for considerable trouble if anything goes wrong.

This strategy is thus very risky, particularly in the period before an agreement has been reached. The seller's marketing expenses are often very high already in the early phases of a business relationship due to all the contacts that need to be cultivated with buyers to inform and influence them, and to create trust. Also, the costs for preparing the bids on which future agreements are based are normally considerable. Since each solution to a customer problem is specific, most of the expenses are wasted if the contract is won by a competitor. The degree of linkage specificity is high and switching costs are high. It is thus important that the parties protect themselves from the dangers inherent in this type of business dealings. Since the degree of substitution of the relationship is low, this is usually done within the relationship by building a mutual balance of gradual commitment of resources. This is done in terms of the relationship. The decisive judgments for both parties concern the degree of commitment to be made in each phase of the process which leads to a contract. After an agreement has been reached, commitment increases further as the project is installed, operated, and maintained. With a contract in hand this is less risky. The transactions here are thus characterized by a low degree of substitution, strong investment in linkage-specific assets, a long-term character (though with considerable irregularity), and the view that marketing requires the selling of ideas. This places very special demands on the seller.

The process just described is largely valid too when relationships are established within a bilateral governance form. In the habitual stage, for example, the parties have become so united that other parties are excluded, competitiveness is maintained, and transaction costs are kept low through a certain standardization of procedures which serves as a barrier to competitors. It is essential not to lose this competitiveness by letting the relationship become set in a fixed mode. Flexibility is maintained by adaptations to a changing context.

The Competitive Strategy

The competitive strategy, the second part of a marketing strategy, is more closely related to horizontal competition. The seller offers a technical solution to a buyer's problem. It is contained in a package consisting of various offers, for example of hardware in the form of products of a certain quality and software such as service, transfer of know-how, and financing. The package is delivered within a specified time at a specified price. It is modeled in such a way that it differentiates itself favorably from the competitors' offers. Since the package does not sell itself, it is marketed mainly on the basis of a linkage strategy. Selling information is transferred personally and social influence is created. The seller may sometimes get support from a third party such as a trade council or government.

Governance forms create the conditions under which a competitive strategy is performed. Firms acting within the constraints of market governance use price as the main strategic component. In the other governance forms price competition is not the dominant component, although it is still important: In bilateral governance it is simply one of the factors of importance; in trilateral governance it is one of the matters agreed to in the formal contract, but is usually subordinated to the product itself and its attributes. To be competitive in the marketing of large projects in SEA, a TNC tends to need a strong local organization, financial strength, reference projects, and high technical ability.

Price

An important function of the governance form in question is to establish norms for the functioning of price in an institutional sense. In market governance price is by far the most important aspect. In neoclassical theory, price as an *ex post* equilibrium price is in fact the only norm. As Hodgson (1988, 186) observed:

> It is presumed to be formed after an extensive process of market adjustment and price signalling in logical rather than historical time. In contrast, although the kind of price norm discussed here may be affected by day-to-day prices, in another respect it exists *ex ante*, embedded in institutions and the expectations of individuals, and thus bears upon current prices in historical time.
>
> It is important to note that price norms acquire a moral dimension in the eyes of the purchaser, which further helps to reinforce them in the market.

Considerable resources are spent in exchanging products to fulfill the needs and requirements of society. Here prices and products are not seen as given as they are in neoclassical economic theory. Rather, prices and products are viewed as first negotiated and then determined (product and factor prices). Market actors are considered as price makers and not as price takers. This is a very difficult and risky task, the outcome of which determines the distribution of economic benefits between the parties. Prices are considered not only to be discovered but also to be influenced by the parties, as well as to be dependent upon the resource situation (or economic power) of the buyer and seller and of their competitors (e.g., oligopoly).

How are prices fixed in various governance forms? In market governance, prices are determined by the market, and the market actors are mainly price takers. In bilateral and trilateral governance, on the other hand, the parties are the chief price makers. In the presence of either of the latter two governance forms, it is expensive to determine the price of a product, both for the buyer and for the seller. The seller, for instance, collects information about the segment of the market of relevance (customers and competitors) as well as concerning one's own activities, such as objectives, strategies and various costs relevant to the product. The own-activities aspect determines how the internal transfer price system is to be organized. In market governance the price provides a good approximation of efficient resource allocation. It is often the case, however, that this correspondence between price and resource allocation is more difficult to maintain in bilateral and trilateral governance. Nevertheless, this does not preclude companies' stressing efficiency goals. As already discussed, a company's primary aim is to stay competitive through efficiency. It is in the interest of sellers then to offer cost-effective products and for buyers to purchase such products in order to be competitive in the markets in which they sell their end products. Consequently, the resources spent are recovered through price, at least in the long run. Price is the main approximation of an efficient allocation of resources between the parties to a linkage and between competing parties in the environment. Although it may not be possible to determine allocative efficiency for society as a whole, firms are supposed to act in accordance with efficiency goals. Price reflects, therefore, not simply production costs but also transaction costs. The lower the transaction costs, that is, the costs for discovering, fixing, and enforcing prices, the less resources used to fulfill needs, the lower the prices, and the more competitive a company. This is "the bottom-line" in pricing.

As already indicated, the connection between price and allocative efficiency varies with the institutional context. The connection is more straightforward in the market governance form and more complex in the bilateral governance form. Pressure from the environment for linkages to be efficient is stronger in market governance than in the other governance forms. Even if it always is in the interest of the parties to transact efficiently, since it increases their competitive strength, they are less constrained to do so by the market environment in bilateral and trilateral governance. It is particularly difficult to fix prices altogether for very heterogeneous products which are sold in risky markets under conditions where other basic institutions, such as regimes of appropriability, are weak or deficient.

From what has been said, one can conclude that, in a sense, price is the most important component of competition in all governance forms, since it reflects the resource allocation between the parties and competitors involved. It can be seen as the very basis of competitiveness. However, as already indicated, it is particularly difficult in the case of bilateral and trilateral governance to establish the relation between price, transaction costs, linkage efficiency, and competitiveness. Pricing is more complex in these forms. First, it is difficult to calculate prices and to know whether they reflect an efficient resource allocation between parties. Second, non-price competition is also important. This type of competition largely dominates marketing of advanced industrial products by the TNCs studied, relegating price to the background there in competitive strategy. Even if price agreements finally determine the resource allocation between the parties, quality, service, and other competitive factors may be stressed to a greater extent by buyers and sellers. Although the two are related to each other, it is vital to distinguish between price as a competitive factor and price as a reflection of resource flow. Complexity of needs makes it difficult to price an offer. The price must be related, for example, to a certain quality and service level as well as to delivery time. These parameters are determined collectively within the framework of the linkage, as the information sought and obtained and the bargaining and agreement enforcement necessary enter into its determination. Price is one of various factors determined through such a process. However, it holds a unique position in reflecting the resources consumed and their allocation between parties.

The high uncertainty involved in bilateral and trilateral governance and the ensuing problems of fixing prices, making them less transparent, can be taken advantage of by an opportunistic seller. The buyer can be influenced to buy a product that is less than suitable to actual or future needs. Similar problematical results could come about through neither party having sufficient knowledge concerning the needs in question and the potential alternatives in the market. The non-transparent role of price as a competitive factor under such conditions may thus result in lower allocative efficiency.

Price tends to be evaluated in terms of expectations and promises regarding such matters as quality, service, and delivery. Exchange only takes place if price is acceptable to both parties. In trilateral governance, its role varies with the substage of the linking process. At the first substages it functions in the background, since activities mostly concern establishing other aspects of an agreement, mainly the product, that is, determining the needs of the buyer and how these can be fulfilled.

First-Mover Advantages

As already observed, a crucial part of strategy is to find the most efficient mix and sequence of linkages over time in order to achieve first-mover advantages. "The basic phenomenon is this: Winners of initial contracts acquire, in a learning-by-doing fashion, nontrivial information advantages over nonwinners. Consequently, even though large-numbers competition may have been feasible at the time the initial award was made, parity no longer holds at the contract renewal interval" (Williamson 1975, 34). Such a conception has been transformed in this book into a strategic marketing concept. First-mover advantages may be achieved either through information linkage, in which the customer is influenced by and dependent upon information from the seller, or through social linkage, with the customer becoming socially committed to the seller. In addition, such advantages may be achieved through product or financial links, in which the customer becomes dependent upon the seller's products and financing. Specific linkage mixtures may be found to be particularly advantageous at various stages in the building of the relationship. In the marketing of projects, for example, social and information linkages are very critical at the initial establishment phases.

A first-mover advantage can also be achieved through a competitive offer. The strategy here involves a combination of a linkage mix and a competitive mix, which are interconnected. The linkage mix creates a framework for the transfer of the competitive mix, but is also influenced by what the TNC as a seller can offer. An advanced technical solution far above the customer's present capabilities, for instance, requires a more long-term accumulation of information and social contact networks for transferring know-how. Likewise, if financing is an important marketing element in the package, a financial linkage will develop.

THE LINKING PROCESS

Information costs, bargaining costs, and enforcement costs are the main transaction costs affecting how a solution to a customer need is marketed. The first stage of the marketing process to bridge this gap between demand and supply is called the establishment stage. This stage, which ends with the implementation of an agreement, is followed by what is termed the habitual stage. Here the relationship has already been established and routinized, and appropriate habits have been developed. The relationship then functions in accordance with certain customs. In the earlier part of the establishment stage information costs predominate, as do bargaining costs in the later part. Enforcement costs are predominant in the habitual stage. The establishment and habitual stages are common for bilateral and trilateral governance. The two governance forms differ in certain characteristics of the respective stage, such as their relative importance and their duration. Establishing a relationship takes a longer time in trilateral than in bilateral governance, the habitual stage predominating in the latter form.

The major aim in marketing is to reduce the costs of the various transactions during the linking process generally. At the same time, there is a trade-off between the three main types of transaction costs. High information costs in the initial stages could be compensated, for example, by lower bargaining costs at a later stage.[2] The economics of transacting vary as the process develops. Both the greater knowledge of each other which the parties gain and the transacting process itself make it increasingly possible to organize transactions efficiently. Through bilateral governance in particular, the process described above can be simplified considerably and transaction costs saved, and thus the linking process will become routinized.

As more and more transactions are completed, the seller is able to reorganize them so that they become increasingly efficient. As discussed elsewhere (Jansson 1992), some of the companies studied have reorganized their sales units in SEA to take advantage of accumulated experience and to increase sales. With the stabilization of transactions, knowledge of prices and of other conditions have tended to reach a steady state. Parties have learned much about each other and about competitors, and this in turn makes it expensive to change prices. Interest in prices has gradually decreased and price competition has been replaced by non-price competition.

A common way of economizing on transaction costs when introducing a new product in Southeast Asian markets is to take needs more or less for granted, assuming them to be essentially the same as in the "old" markets of Western Europe. If the company assumes that the requirements for discovery of prices in the region are low, this will reduce transaction costs in this respect. If the original assumption is incorrect, however, and reflects poor local responsiveness, a long-term competitive position may not be achieved and profitability will be affected. This is because the marketing situation was not properly analyzed. Through necessary adaptations thus having been delayed, later transaction costs may be considerably greater than would have been necessary with better adaptation to local conditions.

STRATEGIC MARKETING PROFILES

A marketing strategy describes a certain action profile of a company. Governance forms and marketing situations describe the conditions for such action. To be able to take advantage of the prevailing conditions and to act, resources are needed. Thus, a company's action profile can depend very much on the types of resources or assets it has available, or the types of products it sells. To clarify these matters, five basic strategic marketing profiles are presented. These describe the specific strategy-relevant assets found at a company, such as knowledge, skills, or equipment (soft/hardware). To decide on a strategy and implement it in a certain market environment, specific types of resources are needed, for instance detailed knowledge about a few customers or a general knowledge about many customers. The strategic profiles considered have to do with the company's ability to handle various types of linkages. They are thus to be found at a kind of intersection of strategy and structure.

The grouping of the profiles, clarified in Figure 4.3, is based on a number of distinctions made within the framework of the conception of marketing employed here. First, the distinction is made between two main types of solutions, whether the resources of the firm are directed at satisfying specific customer needs or more

FIGURE 4.3. Strategic marketing and manufacturing profiles.

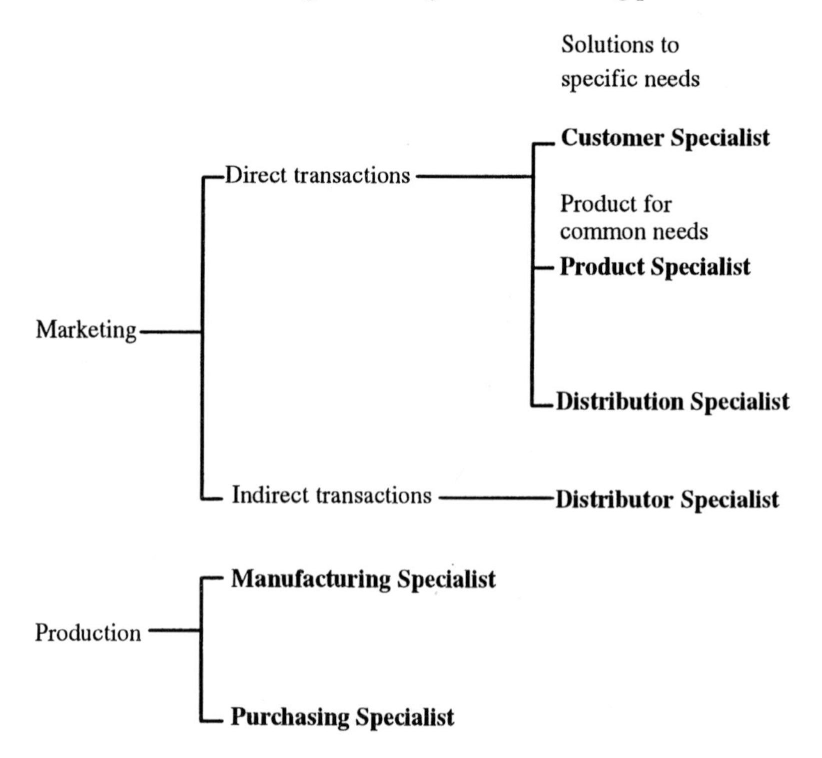

general ones. To bridge the gap between needs and solutions in the first case, solutions are adapted closely to the needs of the individual customer. Information there is collected both about specific customer needs and about needs of customers in general. Possible solutions are found and evaluated, and are later negotiated and enforced. Such activities are directed at specific customers. This strategic profile is called a customer specialist. In contrast, when activities concern the solving of common customer needs, actions tend to be directed at and converge on the product instead of the customer. The strategic profile this involves, the focus of which is the product in general and not more specific problem solutions, is that of a product specialist.

Second, the distinction is made between two different ways of bridging the gap between needs and solutions: through direct transactions between the buyer and the seller, on the one hand, or indirectly through a third party, on the other. The profiles of product and customer specialists are both oriented toward direct contacts. In contrast, where transactions with customers are indirect, distributors come into focus. A strategic profile directed at indirectly bridging the needs-solutions gap, and in which transactions with intermediaries are more important than transactions with final customers, is called a distributor specialist. A profile, on the other hand, involving specialization at being an intermediary or linking-pin, is called a distribution specialist. All of the four marketing profiles referred to thus far are found among the sales subsidiaries that were studied.

Third, the distinction is made between marketing and manufacturing. As was emphasized in Chapter One, manufacturing is important from a marketing point of view, since the efficient production of a solution/product is a prerequisite for good marketing results. A profile oriented towards production, as was the case in some TNCs which considered local production to be an important factor in competitive strategy, represents that of the manufacturing specialist. Sales companies in the conventional sense represent, in turn, purchasing specialists. Manufacturing companies, to be sure, may be involved in both roles. If a company is classified, on the other hand, as a manufacturing specialist only, it has no capacity for marketing, and may serve as a subcontractor of machine capacity. No such company, in fact, was found in the present study. Only a small fraction of the activities of some few TNCs in the region fall under that heading. Certain local companies, trading companies, for example, which specialize in sourcing products for a TNC, could be considered pure purchasing specialists of a sort. The activities of such companies are not discussed here.

Each of the strategic profiles described involves the refined use of certain strategic categories of resources. When the resources of a firm and its use of them are classified according to this scheme, a particular profile or a mix of profiles is obtained. Such a mix can be further classified into various aspects which are dominant and subordinate, respectively.

Customer Specialist

Problem solving directly oriented to the specific needs of certain customers is the main feature of the customer specialist's strategic profile. The emphasis may well be on software aspects, designing these and adapting them or other technically advanced products to individual customer needs. Engineering is the key concept here.[3] Activities denoted by this profile are characterized by long lead times, high sales costs, broad contact nets, and long-term relationships. Production is subordinated to marketing and must be flexible in order to make possible the production of the specific solutions needed. The scope of the problem-solving capacity available varies with the numbers and kinds of customer needs. Customer specialist capabilities tend to emerge through continual contacts with customers, and through learning about and adapting to their special and often changing needs. A deliberate planning of the company's operations is difficult, making frequent contacts with customers particularly important.

The degree of linkage specificity is high. Physical and human asset specificity, as well as degree of impactedness, are usually high due to the specialized nature of the products sold. This also makes the costs of switching between customers substantial.

Competition in the market here is need-positioned, takes place between different networks of specialized linkages, and is directly oriented toward specific customers and the satisfaction of their special needs. Bilateral governance is typical. Trilateral governance in this context involves competition for project contracts.

Product Specialist

Companies which sell technically advanced industrial products with little or no adaptation to the special needs of individual customers are classified as product specialists. Marketing activities here concentrate on the product, which is sold to customers with a common need for functionally high and consistent quality. A high degree of technical capacity is essential. Technical capacities should be high throughout the company, in marketing as well as in manufacturing and purchasing. This expertise, together with skillful coordination of operations in all three areas, is necessary if an efficient strategy is to be achieved and first-mover advantages are to be maintained. Product development is the key to successful operations. Economies of scale are important in all operations. At the same time, the need for flexibility in marketing and production is less than for the customer specialist. Linkage specificity is low to moderate, and asset and product specificity are low. The degree of impactedness can be anywhere from low to moderate, depending upon how the customer interface is organized. Stability is more important than change. The organization of the product specialist is reminiscent of machine bureaucracy. Long, stable periods alternate with short periods of change, and the company follows a deliberate strategy that is maintained for an extended period. The product specialist, in contrast to the customer specialist, lives by exploiting a strategy rather than by changing it.[4] Competition is product-positioned and oriented toward customers in general. Products are offered and competition takes place between a standardized system of linkages.

Distribution Specialist

Sometimes the high costs of establishing a distribution network may be justified provided that matters of pure distribution are important and that the distribution required is not too intensive. The number of customers, industries, and product varieties must be sufficient to justify the company's having its own distribution network. Functions other than distribution may be important too, especially those involving direct contacts with customers. Distribution here can be characterized as more selective and exclusive than intensive.

When a company establishes itself as a distribution specialist, it also establishes its own sales company and becomes a product and customer specialist as well. If the company had been represented on the market by an external party, the transformation to distribution specialist means internalizing distribution. This is to be distinguished from a company that is specialized in sales internalizing its agents, and thus becoming only a product or customer specialist. Distribution, product, and customer specialists all aim to integrate forward to their customers so as to have direct contact with them. The main difference between the three is in which aspects of marketing are emphasized. Whereas product and customer specialists concentrate on sales and service, respectively, distribution specialists place emphasis upon the distribution aspect, for example through assembling a broad assortment of goods, and concentrating on their physical supply, transportation, and storage.

Distributor Specialist

Many products are not economical to sell and/or distribute directly to customers. For the distributor specialist, transactions are externalized to distributors. It is important to distinguish distributors from agents. Distributors are middlemen who buy and sell industrial goods but have distribution as their chief marketing function. Besides undertaking sales and warehousing, they may also offer customers service and financing. Agents, in contrast, only represent specific sellers, usually only one seller or a small number of these, and they specialize in some particular geographical area and in often only a small number of industries. Although agents may stock the product, they concentrate mainly on selling it. This is one alternative to a company having its own sales subsidiary.

It is economical to externalize transactions to distributors when selling small quantities of a broad variety of products often of rather simple technical character. Transaction costs are lower when these products are sold through an intermediary rather than the company itself serving as a distributor. A distribution network which is already established may be utilized. The costs of obtaining information about customers and competitors are lower when such a network is utilized here. Bargaining with intermediaries is less expensive than bargaining with the customers directly and the costs of controlling the prices of the intermediaries is lower than that of enforcing prices directly with customers or through one's own distribution network. The cost of obtaining information about the market and the intermediaries in order to control the prices of distributors is one of the main transaction costs for the distributor specialist. These costs are lower for simple standardized products, since prices for these are mainly controlled by the market. These are also the main types of transactions which were externalized by the TNCs which were studied.

The availability of products in the various parts of the industrial distribution network is a critical competitive factor for the distributor specialist. It is vital that the right product be delivered to the right customer at the right time. For customers to be satisfied, delivery systems, inventory management, service, maintenance, and sales must function well and warehouses must be located advantageously. Ideally, the distributor specialist should have thorough knowledge of the needs of customers. Acquiring such knowledge directly would necessitate direct transactions with customers, however. Thus, there is a certain pressure on the distributor specialist to try to gain greater control of the marketing process. At the same time, gaining such control could require high transaction costs and greater linkage specificity than the distributor specialist normally has. Nevertheless, at some stage it may appear more advantageous to the distributor specialist to integrate forward and buy distributors and/or establish its own distribution system. Since this solution involves considerable expense, the expected gain in efficiency should be considerable for this to be selected. Another alternative is that the company partially integrate forward to set up its own sales company to participate in and control the distribution network on the spot. In this way the TNC moves closer to customers. It is still basically a distributor specialist, since distributors still supply the products. The distinction can nevertheless be made between this type of distributor specialist for which conditions are such that both distributors and the distributor specialist have direct transactions with customers, and the distributor specialist more purely concerned with distribution. A TNC invests more in the distribution network in the former case than in the latter.

CONCLUSIONS

Linkage strategies and competitive strategies and how they are related to first-mover advantages were discussed in this chapter. The major stages of the linkage process were related to three major types of transaction costs. Strategies were shown to follow an economic logic that could be expressed in terms of a transaction-cost model. Linkage specificity was shown to be a key concept in SEA. In this context, the distinction was made between asset specificity, product specificity, degree of impactedness, and degree of substitutability. Basic to all of this is the goal of efficiency, achieved through reduction in transaction costs. Strategy and organization are conceived within the broader framework of governance forms. Four strategic marketing profiles were found to be important with the present context, those of the customer specialist, product specialist, distribution specialist, and distributor specialist.

Although the role of price varies with the governance form, it is the main information carrier and control mechanism in all of them. Price is of critical importance then for efficient resource allocation and is the crux of competitive strategy.

NOTES

1. This situation, in which the initially winning bidder has an advantage over the nonwinners, is also termed "the fundamental transformation."
2. This process of transacting is similar to one described by Commons: "At the negotiation stage the future values of a transaction and the particular working rules which will govern future conduct are determined. In the transaction stage, commitments for future action are transformed into an explicit executable agreement or contract. The final administrative stage is the actual execution of a transaction" (Leblebici 1985, 106-7).
3. This profile is also typical of international technical consultancy firms.
4. Compare Mintzberg (1987).

Chapter Five

Linkage Strategies

This chapter concerns how gaps are bridged between the company and its customers. A wide variety of different linkage strategies are available. The transaction costs of the exchange are a central factor in the choice of a linkage strategy. Many different aspects of this strategy may vary. One aspect concerns what is transacted. This may be, for example, any of the following: resources, such as the products, service and know-how; finances; information (communication); affection (social exchange); or norms (expectations). Another aspect involves the question of whether transactions are direct or indirect, in the latter case, for example, going by way of such intermediaries as distributors, consultants, or brokers. A third aspect concerns the manner in which transactions are conducted, for example personally, by word of mouth, by telephone, by personal letter or by facsimile, or impersonally by ads, pamphlets, form letters, and the like. A fourth aspect is time. Not only do transactions take time to complete, but their relations to one another can also change over time, their character sometimes differing, for example, between the beginning and the end of the transaction process.

Linkage strategies are often formulated and executed in conjunction with competitive strategies; the linkage strategy can pertain, for example, to how different resource flows such as those of product, service, or training should be combined in the competitive strategy. An offer of a competitive price, for instance, presupposes that the linkages are efficient so that transaction costs can be low. Linkage strategies are partly a function of the environments in which they take place (e.g., the marketing situation, the institutions, or the governance forms involved). They are also constrained by the company's strategic marketing profile, whether it is a product specialist or a customer specialist. Throughout the chapter, linkage strategies are analyzed in terms of these two specific strategic marketing profiles. One aspect of the environment is particularly important, the marketing situation, or more precisely certain associated characteristics of the customers, in that it helps to elucidate the implications of the two major marketing profiles. This is developed in the first section of the chapter. In the next section, the segmentation of customers is discussed. In the final section, which covers buying behavior, a further environmental factor, that of business culture, is considered. This section helps make clear why social exchange is such an important part of Southeast Asian linkage strategies; it is a critical element in marketing strategies there. Methods by which social linkage is established and maintained through contact nets is emphasized throughout the chapter.

LINKAGE STRATEGY AND STRATEGIC MARKETING PROFILES

A linkage strategy is a kind of tying-up strategy in which the strength of the bond varies with the strategic marketing profile. The stronger the bond, the more dependent the parties are on each other. Transaction costs for switching to alternative suppliers/customers are then high. This implies a high degree of specificity and a low substitutability of the linkage. Customer specialists that sell to relatively few customers are characterized by a high degree of tying the parties with which they establish linkages. In contrast product specialists selling to many customers do not have such strong bonds with buyers. In the former case linkages are concentrated and specialized to a few customers, whereas in the latter case they are more spread out and standardized to a wide variety of customers. As already noted, the linkage strategies are quite different under these opposing circumstances.

Such differences in linkage strategies depending upon strategic marketing profiles and customer attributes are evident in Table 5.1. The grouping of firms there is based on two dimensions: number of customers and character of the products sold. The last dimension is important to linkage specificity as this relates to strategic marketing profiles. Products adapted to the specific needs of the individual

TABLE 5.1. Customer specialists and product specialists.

NUMBER OF CUSTOMERS	DEGREE OF CUSTOMER ADAPTATION	
	SPECIAL PRODUCTS	**STANDARD PRODUCTS**
FEW CUSTOMERS	***Narrow Customer Specialist***	***Narrow Product Specialist***
	Telecom (MJE) Paving Systems (MJE) Food Equipment (MJE) Tooltec (PE) Indpow (MJE)	Pacmat (MJE) Conmine (MJE/MIE) Explo (MRO) Indpow (MJE/MIE/COM)
MANY CUSTOMERS	***Broad Customer Specialist***	***Broad Product Specialist***
	Weldprod (MJE/MIE) Wear and Tear (MRO) Indpow (MJE/MIE/COM Food Equipment (MIE/COM) Drives (MIE)	Weldprod (MIE/MRO) Wear and Tear (COM/MRO) Indpow (MIE/COM/MRO) Food Equipment (COM/MRO) Tooltec (MIE/PE/MRO) Conmine (MIE/MRO) Explo (MRO)

Abbreviations:

COM = Components and subassemblies
MJE = Major equipment
MTE = Minor equipment
MRO = Maintenance, Repair, and Operations items
PE = Processed materials

customer are termed customer-adapted or special products. Uniform products catering to the needs common to a large number of customers are termed standard products. A scale could be developed with highly specialized products at the one end and highly standardized products at the other. This is not necessary here since it is sufficient to distinguish between two main groups: customer specialists and product specialists. First, the data collected does not permit such refinement. Second, we are mainly interested in examining relations between certain crude categories. We thus have to make certain simplifications. A special product, for instance, is designed and manufactured for a specific customer need, either directly through a tailor-made process or by assembling different standard, prefabricated modules. Although special steels, machine tools and compressors which are manufactured in many standard varieties can be seen as intermediary cases between special products and standard ones, they are classified here as standard products which to a certain degree can be adapted to various broad customer requirements. Both standard and special products can also be specially adapted to different technical levels, for example to a lower technical level in product needs within LICs. Such adaptations common to both categories fall outside of the basic grouping here. The customer dimension is also rather rudimentary, as it is divided into only two classes: few customers and many. Companies that sell to many customers tend to have a broad strategic profile, whereas those that sell to few tend to have a narrow profile.

Except for the broad product specialist's standardizing of both products and customer needs, all four strategic marketing profiles–narrow and broad customer as well as product specialists–involve a specialization of some kind. Customer specialists adapt their products, various marketing functions, and different parts of their organizations to individual customers, whereas narrow product specialists concentrate on a limited number of customers.

In Chapter One the industrial products marketed by the European TNCs studied were classified into seven categories (Table 1.3). These products are sold within quite differing sectors and industries, as is evident in Table 1.4. Table 5.1 indicates that most of the TNCs have rather concentrated strategic marketing profiles. The products of most of them are then found in only one or two of the cells. However, there are certain companies with more mixed and complex profiles which are found in all four cells, for example Indpow. Food Equipment lacks any narrow product specialist profile. Weldprod and Wear and Tear are classified as both broad product specialists and broad customer specialists. Drives, on the other hand, has a more concentrated profile, that of the broad customer specialist. Conmine, Tooltec, and Explo are mainly broad product specialists, but with a minor ingredient of the narrow product specialist or customer specialist when products are marketed to projects. Pacmat is classified entirely as a narrow product specialist. Telecom and Paving Systems can be viewed as narrow customer specialists. Indcomp does not belong in these groupings at all, since it is classified as a distributor specialist. The linkage strategy of that strategic marketing profile, which is also found in certain other companies (e.g., Tooltec), is described at the end of Chapter Six. Specmat, lastly, is the only firm here which is classified as a distribution specialist.

STRATEGIC MARKETING PROFILES AND SEGMENTATION

Product specialists with a customer base of few but large customers tend to differentiate between the needs of their various customers more strongly than do broad product specialists. Pacmat has a very concentrated customer structure, having only a few customers, most of them of considerable size, and attempting to get to know each customer very well so that it can implement highly individualized linkage strategies. Pacmat involves itself very much in its customers, even helping them to market their products. Since the customers themselves compete with one another, Pacmat's relations with them, if they are to be successful, must be based on a high degree of trust. In countries of multiple cultures such as Thailand and Malaysia the cultural dimension often tends to segment the market, too. In Thailand, for instance, Pacmat's local customers are segmented into five separate groups of Chinese and Thai companies, whereas Pacmat's international customers operating in the country are classified into two major groups. Customer specialists with a limited number of customers in SEA tend to divide their customers into different groups and to adapt their contact nets to each of them. Telecom, for example, markets most of its major equipment to only one customer in each of the SEA countries. In most cases the customer is a very large organization subdivided into many different units. Broad product specialists and customer specialists, in turn, although they do segment their customers, particularly those of large size, do not carry out segmentation to the same degree. The grouping tends to be less detailed and to be based partly on the type of industry involved and partly on the home country of the customer. Regarding the latter aspect, a major distinction tends to be made between local firms and TNCs, since the two differ very much in buying behavior, displaying differences in their demand for high-quality products, and price-related behavior. Customers are commonly separated into three groups: price conscious buyers, quality conscious buyers, and a group in-between. Local customers are often grouped by whether they are private Chinese companies or public companies. A segmentation of TNCs according to their nationality in terms of the categories *European*, *U.S.*, and *Japanese* companies is common. Sometimes the categories *Korean companies* and *Taiwanese companies* are also employed. There is a definite connection between the purchasing behavior of these TNCs and the country of origin.

STRATEGIC MARKETING PROFILES AND CONTACT NETS

The more specialized the products, in terms of being adapted to the needs of the individual customer, and the smaller the number of customers, the broader and more intensive the contact nets formed by individual seller/buyer linkages. For narrow customer specialists social linkages are particularly critical. A high degree of linkage specificity involves a high degree of social impactedness as well. Contact nets are an operationalization of this. In the case of broad product specialists, in turn, the contact net tends to be both narrower and more shallow. The other two strategic marketing profiles considered are found in-between these two polar cases of narrow customer specialist and broad product specialist. It is difficult to compare in a precise manner these intermediary types of strategic marketing profiles. On the one hand, a broad customer specialist can appear to have rather intensive contacts with its many customers due to the higher need of individual customer adaptation of its products. On the other hand, this may well be surpassed by the closer contacts which narrow product specialists must establish in their sale of standardized products to a few large but differing customers.

The basic relationships described between numbers of customers, types of products sold, and contact nets are probably valid in most markets, and are not special for SEA. The present study does show, nevertheless, that contact nets are a particularly important part of linkage strategies within this geographical area. A main reason for this can be seen in the cultural characteristics of the Chinese, the dominating local business group. Culture is thus an important element in explaining why contact nets are so vital for strategic marketing profiles in SEA. Social linkage there plays a central role.

Common to all the cases studied was the finding that personal relations and social contact nets are fundamental to marketing success. Personal selling was found to be of exceptional weight in comparison with other direct but more impersonal ways of transacting such as by phone, by telex, or by mail, or with more impersonal and indirect media, such as that of advertising. Direct personal contact can thus be seen as the core of linkage strategies. Individuals provide the major means of contact between buyers and sellers and are central to information, social, product, and financial linkages. The competitive strategies of those TNCs studied that were concerned with the transfer of technology and service involved to a high degree direct contacts between employees of the respective companies. As already indicated, the contact patterns varied in breadth and depth. For many of the products sold by the TNCs studied, the contact net was rather broad and involved several company representatives of different management levels. Contacts tended to be special, intensive, and broad. They took time to build, were of a personal character, and could take place at any time of the day or night. They did not necessarily have to do directly with business and they were maintained even if no order was received and there was no particular expectation of receiving orders in the future. It is critical to establish trust in such contacts. However, this does not provide any guarantee of continuous and long-term business. Indeed, when salespeople leave a company, they may take their customers with them. The importance of social linkages makes sellers invest considerable resources in the creation and maintenance of contact nets. Otherwise no first-mover advantage can be achieved.

LINKAGE STRATEGY OF A NARROW PRODUCT SPECIALIST

The linkage strategy of a narrow product specialist with its concentration of sales to a limited number of customers is well illustrated by the Pacmat case. The main product sold by that company is a processed material which is manufactured there in the area with use of a mass production technology. Major equipment is also sold, for which the processed material represents the main input. The major marketing strategy of this TNC is thus to prepare the ground for the chief product through the sale of the major equipment involved. It is also vital for it to have a facility close to the customer for servicing the equipment.

Pacmat strives for a few strong, locally based customers in each of its various fields of business. The more each customer can sell the more it will purchase Pacmat's products. If Pacmat had too many customers, this would cause too much competition among them. Since Pacmat is their main supplier, this competition would result in low profits for the companies. It would also burden technical service too much. Pacmat involves itself very much in its customers, supplying them with the major equipment and related service, as well as with the main input for this equipment. It also provides them different kinds of assistance in such areas as marketing.

Market Segments

Pacmat divides its customers in Thailand into seven major segments, as illustrated in Figure 5.1. A segmentation very similar to this is also utilized in Malaysia. The same three local Chinese groups and the same two TNC groups are also found there. However, since Malays are more involved in business than the Thais, and the former are Moslems instead of Buddhists, the overall grouping is different there compared with the following segments in Thailand.

1. Small family firms run by a husband, a wife, and their children. They work 20 hours each day and travel only on business. They are totally involved in their companies. Employees come mainly from the same family or clan, or they are recruited from the same village or the same province, and they must be Chinese. The persons working in such a firm are traditionally very Chinese in their outlook.

2. The Chinese tycoon with his firm. This is usually a somewhat larger company. It is run by this entrepreneur/tycoon entirely.

He and his family are not completely assimilated into Thai society since they belong to a first or second generation there. This tycoon is not liked much by other assimilated Chinese or by the Thais. He is often looked upon as a kind of upstart. His own outlook is also traditionally Chinese. This manifests itself, among other things, in his being proud of his original home country, by which is usually meant China before the revolution.

3. The Chinese-Thai type of business group. This is the "big business" part of the economy. The nucleus of a group is often a large bank. Many separate firms of the two former types may be linked with it through the borrowing of money. The Chinese here have lived in Thailand for many generations and are assimilated. They consider themselves to be Thais. They have Thai names and are somewhat aristocratic and conservative. However, in business they are more modern in their outlook than the tycoon, whom they despise and look upon as primitive. They are usually rather westernized and are interested in cooperating with TNCs.

4. Thai firms (The Business Man). These are still very small and are of minor practical interest.

5. The Thai bureaucrat. He acts in a completely different way than a Chinese. Persons employed in Thai cooperatives belong to this group.

The last two categories to be considered are Country Managers (CM) of subsidiaries of TNCs.

6. The old CM, more than 50 years of age. He has lived in the country for a long time and often has a Thai wife. Such a rather well-assimilated Westerner is often called a "Farang." For persons of this category Thai cultural values such as that of seniority are often very important.

7. The expatriate CM. The assignment he has tends to be temporary and to be only one of the many he has had or will have

in his career. No adaptation to local cultural circumstances is necessary for persons in this group.

Contact Nets

Pacmat tries to identify the right persons within each segment, helping it to approach its customers in the right way. When business was established in Thailand, for example, the European CM stationed there had contacts with members of all these groups. As already emphasized, linkage strategies need to be adapted to such local business cultures.

Pacmat built its contact nets differently for each segment. During the first five years the pattern of contacts was as follows: The CM's main contacts were with Managing Directors (MDs) and owners within each of the customer groups. When the CM became a friend of the tycoon, for example, he also obtained authorization and access to the rest of the organization. The CM also maintained good contacts with Western managers at the level next below that of CM within TNCs in the area. His contacts at that time were largely with

FIGURE 5.1. An example of cultural segmentation.

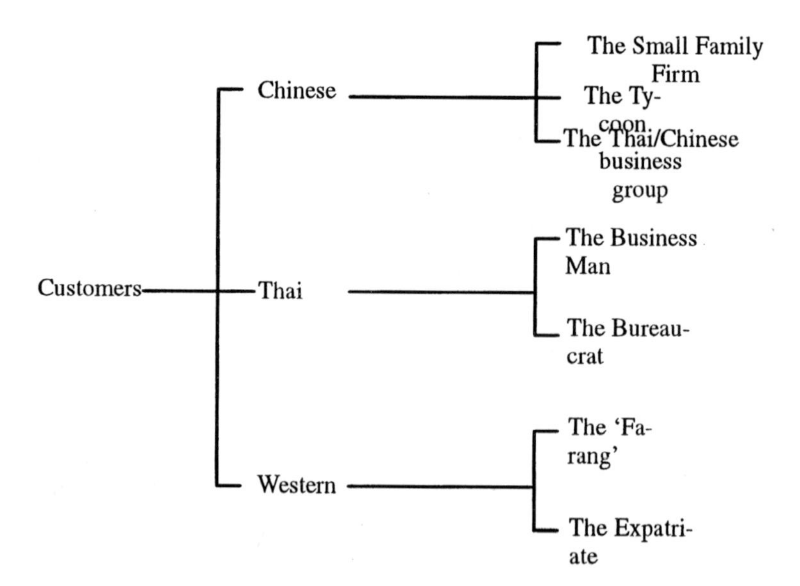

Westerners. The marketing manager of Pacmat, who was Chinese, had few contacts with other TNCs. In his dealings with Chinese firms his contacts were with such persons as product managers one level below that of MD. In the marketing manager's contacts with Chinese-Thai business groups, in contrast, he dealt with persons at the MD level. For maintaining contacts with small Chinese firms, a plan was arranged between the CM and the marketing manager such that one of them played the role of "good guy" and the other the role of "bad guy," mainly to be sure they would be paid. On the whole, contacts with these small business firms were hard to manage.

Pacmat's technical manager had contacts with the customers' technical manager and staff, while service engineers and the materials manager had contacts with the customers' designers and foremen. Persons in finance also had considerable contact with customers in matters of debt-collection. Drivers meeting the customers' drivers also represent an important means of contact. The overall aim was to create as broad a contact interface as possible with the parties of interest. Contacts were established professionally and not through clubs, golfing, or activities of that sort, although means of the latter type became more important in later stages, and even then it was important that one be introduced by an older colleague who already knew the customer.

Establishment of Contact Nets

The pattern just described is valid for establishing contacts which will provide a broad contact interface. It was considered best that in the beginning one talk as little as possible about business matters and instead concentrate on matters of more general interest such as weather, events within the country, etc. Family matters were avoided at this stage and were left to the customer to bring up at a later phase. One general insight this company gained and has been guided by is the importance of having a general knowledge of the host country and of its culture. This makes it easier to discuss a wide variety of different matters raised during the frequent meetings which take place, at the same time as it provides a better understanding of the customer's behavior. It is also vital that one adapt to the differing behavior which typifies the various segments of society. Chinese from segment 3, for example, often have been educated in the West and are therefore more westernized and professional in their business contacts. At the same time, at home and in private they tend to be quite traditional. In contrast, the other two Chinese groups tend in their business behavior to be more traditional.

The initial phase was difficult. Later, when contacts had been established by the CM, other persons, such as the technical manager or the marketing manager, were brought along to be introduced to the customer. During this phase, actual business problems were raised and discussed by the parties. It was essential then to offer the customer something concrete in terms of basic information, results of an investigation, product samples, or a study trip to a reference establishment. Which person from the seller's side was selected to participate in such meetings depended upon the type of problem the customer faced. Hard selling had to be postponed until later.

Establishing a linkage can take a very long time. An extreme example is a Thai customer which placed an order 17 years after the first contact had been made. During the many years prior to that the customer was visited once every year to check on how matters stood. It was considered important that a friendly relation be maintained even if a potential order might be very distant. This is not looked upon from a purely business point of view. Normally it takes between one to five years to establish a business linkage. At the other extreme, an order may sometimes be received without any active marketing having taken place at all, for example when a subsidiary of a TNC calls Pacmat and orders something in accordance with instructions from the parent company.

LINKAGE STRATEGIES OF BROAD PRODUCT SPECIALISTS

Conmine is an example of a broad product specialist in which the contact pattern is less extensive than in the case just described. The contacts involved are close, nevertheless, due to the great importance of rapport. Almost all of Conmine's products, their major and minor equipment and their Maintenance, Repair, and Operation (MRO) items, are sold to customers directly. Turnover is divided equally among these two main product groups. Distributors are only used in a few instances involving rather simple MRO items.

Segments

In Singapore most of Conmine's customers within the construction industry are large TNCs from Japan, Taiwan, and Korea. In Malaysia, in contrast, most of the customers in that industry are small quarries, whereas in the Malaysian manufacturing industry, U.S., Japanese, and European TNCs are the company's main customers, local firms accounting for only a small part of the sales. Most of the company's sales generally are to large and medium-size customers. Recently Conmine began cultivating customers new to the area, which have started to invest in local manufacturing industries. Most of these firms originate in Singapore, Taiwan, Korea, and Hong Kong. The importance of the two main types of customer industries, construction and manufacturing, has alternated with the ups and downs of the business cycle. The construction industry was important in Malaysia until 1985, when a long recession started, which did not end until the end of 1988. An increase in turnover which began in 1987 was mainly the result of recovery which occurred in the manufacturing industry. Sales to large projects, found mainly within the construction industry, have followed this same pattern.

Contact Nets

Selling in SEA is a question of establishing and maintaining relationships. Contact nets are usually established by individual salespeople from the company. In Malaysia, where the company has sales branches in various parts of the country, most of the contacts are handled by salespeople from these local units. Experts from the head office in Kuala Lumpur are also brought along when needed. At the initiative of the head office, experts sometimes participate in initial contacts when certain key customers are involved. In the case of large business deals, the product manager participates as well. The service department often participates quite early too, since the need for after-sales service and spares tends to be discussed already during the initial stages. Hints regarding possible new customers often come from this department. Most contacts concern technical matters, and assistance in such matters is sometimes sought from

Kuala Lumpur. This basic pattern for establishing and maintaining contacts is typical for the company all over the world.

In the case of the manufacturing industry, visits tend in particular to concentrate on production and maintenance engineers. Purchasers there are only seldom contacted. Contact nets in the construction industry are focused on construction engineers and purchasers. Conmine has many and broad contacts with its customers all the way down to the grassroots level. The operator is a key person, with whom contact is maintained, since an operator who likes the product commonly tries to influence company superiors. Linkages with public customers, on the other hand, are directed at the people who write the tender specifications.

It is critical that trust with the customer be created. How this is done differs according to the types of industry. In Malaysia, nevertheless, trust is very difficult to establish generally due to the predominance of certain patterns of thinking. Long-term relationships there tend either not to become established or to be readily misused in efforts to squeeze the seller for a lower price. This can be largely explained in terms of the business culture there, which favors short-term thinking, and a buyer's market which has existed there for many years.

Establishment of Contact Nets

The ease in gaining access to various customer groups differs, varying from easiest to most difficult in the following order: European corporations, local firms, U.S. companies, and lastly Japanese and Korean firms. Most European TNCs know Conmine from other markets, which considerably reduces the need to hard sell. The company's competitive strategy best suits European and North American TNCs, whose contact patterns, as well as manner of purchasing, operating, and maintaining their equipment, are very similar to its own.

In contrast, within the Malaysian engineering industry it is very difficult to gain access to new customers. Many visits are necessary before one is allowed to make an offer. Even more visits are then required in order to get an order. To obtain large orders for major equipment it takes anywhere from three months to three years between the first visit and receipt of the order. For minor equipment the period varies from about six weeks to two months. Major factors affecting this are the urgency of the need, the amount of investment involved, and the complexity of the purchasing process.

How often a customer in Singapore needs to be visited varies. Shipyards, for instance, are visited very frequently, usually once a week and at a minimum once every third month. This is necessary to maintain the relationship and to know when the company is in the market. Such information is also obtained from newspapers, friends, and from other contacts. It is important to take account then of the urgency of the need and of the availability of one's own products in comparison with those of competitors.

In Singapore, customers usually come back again after placing the first order. This is a function of the short distances within such a small country, the rather small number of firms, and the fact that long-term support tends to be provided there through service visits and through visits by sales representatives. This facilitates the establishment of long-term relationships. Since Conmine is already well established in Singapore, the company usually needs about three months for the sale of standard products. In other ASEAN countries the situation differs. In countries where the company is unknown, the period is much longer. If customers are not familiar with the Conmine name, they must be acquainted with the company through its products. Some customers will not buy until other customers have accepted the products. References are thus important. The type of product matters as well. For more complicated business deals such as selling major or minor equipment to large infrastructural projects, several years are required to establish a linkage.

Weldprod is a broad product specialist which sells minor items of standard equipment as well as MRO items to such companies as shipyards, certain types of construction companies, and manufacturers of machinery which fabricate products from steel. About 20 to 25 percent of the turnover is in special equipment. The general experience of this company in their direct selling, which amounts to about 80 percent of the turnover, is that customer contacts are personal and much tighter than in Europe. Brand loyalty, at the same time, is much weaker and customers base their purchasing decisions more on persons than on products. A sales representative who leaves the company thus takes customers to the new company.

What persons it is important to contact varies a great deal between different countries, and depends very much upon the complexity of the product and the structure of the purchasing organization. Purchasers and heads of engineering departments are generally the most important persons to contact.

LINKAGE STRATEGIES OF BROAD CUSTOMER SPECIALISTS

The contact net for Wear and Tear's two major product groups is different. The first product group involves a variety of small standard components, whereas the second involves a variety of different small customized items. Both groups of products are sold to large customers directly and to small customers indirectly. The second product group is of particular interest here since it is sold to many large customers in several ASEAN countries, in such areas as the mining, ceramics, and construction industries. Moreover, this product group is the company's largest in terms of sales and is marketed to customers directly, since it is too technically complex for agents to handle. For this part of its business, Wear and Tear is thus classified as a broad customer specialist. The major competitive strength of the company in this field is its knowledge of how to adapt its products to individual customer needs (although, in terms of the first product group of lesser importance, the company can be categorized as a broad product specialist).

It is difficult to sell a technically advanced and expensive type of product in Asia. Due to high interest rates there, it usually does not pay for a customer to invest in equipment with a life of more than three or four years. The efficiency of production is hardly analyzed at all. Large costs for idle capacity arise from various shortages in the economies of the region. In addition, there is a large element of speculation in rapidly expanding industries. Many firms have been established in such industries and have earned large amounts of money in a short time despite inefficient manufacturing. They later close down the when the economy deteriorates. Such behavior is supported by the structure of local industry, with its small, flexible companies and the local business culture of shortsightedness. "Cost-thinking" is often lacking throughout the entire production process. The company must therefore train customers at many levels, which takes time.

Contact Nets

The time spent with a customer varies considerably, depending in particular upon how well developed relationships are, and on whether it is the question of an established or a potential customer. Other factors which influence the time spent are the size and the location of the customer. The largest customers are usually contacted by the MD, whereas two salesmen, as a rule, handle smaller customers. Initially, the MD visits engineers who are responsible for the customer's operations, design, and maintenance to persuade them of the virtues of the product. The next step of the MD is to make the highly important contact with the purchasing department. The customer's MD is also visited, particularly in the case of smaller companies. Primarily technical matters such as how the product can be adapted to the needs of the customer are discussed with engineers. At a later stage, price and terms of payment are debated. Service engineers from the company also come along to participate in the discussions, concentrating on machine operators initially, and later continuing upward in the hierarchy. Since the MD started from the top and continued downward, the representatives of the seller cover all important levels in the customer's purchasing organization. It is considered essential to market the product at all relevant levels. If one person is missed, the whole business deal can be missed as well. Contacts are in most cases personal. It is not enough to send pamphlets followed by telephone calls. Wear and Tear makes on the average eight visits over a period of three years in order to get an order for this type of product. Person-to-person contacts are important throughout the world, but they are particularly important in SEA.

There are many similarities between the marketing of Wear and Tear and Drives in the way in which contact nets are built and maintained. For Drives, contact nets are the focus of marketing. The company's competitive strategy stresses its high product quality in order to motivate price levels above those of competitors. This requires an established contact net. Such a strategy is preferable to engaging in price competition, particularly where the equipment may have to be adapted to the customer by simplifying it technically.

Wear and Tear seems to develop its contact nets in a more systematic manner than does Drives, perhaps largely because of its greater experience in East Asian markets. The type of customized minor equipment which Drives offers is still at the growth stage in terms of sales, as it gradually adds new countries and industries. The company started selling to raw material industries such as those producing sugar or palm oil, and it later expanded to the chemical, petroleum-based, paper-and-pulp and steel industries of increasing importance in the industrial development of the area. As was noted in Chapter One, Drives tends to concentrate on marketing two main types of equipment in SEA, and in Northeast Asia and Australia/New Zealand as well. Due to their small sales volumes in any given geographical area, both companies are dependent on agent representation. Although agents are responsible for contacts with customers, the two companies also take over certain functions from them, mainly those of advanced technical selling and after-sales service, which require direct contact between the manufacturer and the user. For Drives this is particularly the case in the current establishment phase. Volumes are still small and the company is heavily engaged in training its agents and its customers. Agents receive a commission for their sales of equipment, whereas they sell and buy spare parts independently. For the latter business, agents operate more as industrial distributors.

A specific marketing problem encountered by Drives is that its business in Asia is partly a result of agreements made elsewhere. This problem is found in particular in connection with the sale of products to projects. Which equipment should be bought and used in SEA may be decided at the headquarters of the TNC which may be located, for example, in the U.S. A requirement which the purchaser makes may be that the seller be represented at the location where the equipment is used so as to install it and provide after-sales service during its operation. This tends to increase the number of units of the TNC which are involved in selling the equipment. After-sales service, on the other hand, is much more difficult to use as a competitive tool in selling to local industry. Drives' market situation is extraordinary in the absence of Japanese competition and in that there are few Japanese customers.

PURCHASING BEHAVIOR

At several points in this chapter various problematical or noteworthy aspects of the purchasing behavior of customers have been mentioned: customers have difficulties in defining their needs beforehand, decision processes are complicated, decisions are of low technical sophistication, and there are strong social and cultural influences, the latter having such effects as shortsightedness in purchasing. In later chapters other qualities such as price sensitiveness and haggling while purchasing will be stressed. Thus far it has been primarily the consequences of purchasing behavior for linkage strategies that have been discussed. In this section purchasing behavior will be examined in greater detail.

One broad product specialist, Conmine, has extensive experience in selling to the Malaysian engineering industry. Some 80 percent of its turnover has been sold to this industry within recent years, mostly to local units of TNCs. Since 90 percent of the purchasers within this industry are Chinese, purchasing behavior in the local companies and in the TNCs is similar. On the whole, the purchasing practices of the TNCs' local subsidiaries are well developed. The level of capability of the purchasers in the engineering industry has improved over the years. Since late in the 1980s their capacity has improved to the point where they can now evaluate quite technically advanced products in a highly capable way. It is typical for the purchasing organizations that those who participate in the procuring of a product are evaluated in terms of how much they contribute to reducing the price of the product. The seller's MD is even sometimes asked to declare in different phases of negotiations how high a discount a certain purchaser has obtained. Such a policy is extreme, however. Normally it is enough that purchasers show their superiors what has been achieved in revised offers as compared with earlier offers. The contract is also signed by several managers responsible for the different departments involved in the purchase. For Conmine, it is common to have four signatories: the facility manager, the purchasing manager, the financial manager, and the MD.

Most business deals are negotiated. This is particularly true for this company, which sells both major and minor equipment. Even spare parts are sold through offers. In only one out of 25 cases is a product bought directly in accordance with a quoted list price. Business is therefore tiresome and very bureaucratic. Most purchasers are experienced and know how to bargain and to press prices.

Bargaining is very much a part of the mental makeup of the Chinese. Everything is negotiable. There are several reasons for this. One is that the Chinese have been traders for centuries and still are. They dominate trade in SEA and many of the industrial firms and banks in SEA evolved through commerce. Many Chinese are thus born into this tradition, which makes them very deal-oriented and cash-prone. Another factor behind bargaining is the importance of social relations. Lengthy negotiations are a good way to get to know the other party, and to learn whether he can be trusted. A third factor has to do with "face." A rule of business is never let the other party "lose face." A purchaser will "lose face" (pride) toward his superiors if a price which was agreed upon is changed. This dilemma, although difficult, can be solved by including stipulations in the contract which allow for price changes.

Food Equipment is another broad product specialist which is active in this ASEAN country. The company mainly sells minor equipment to indigenous industry. The purchasing capacity of customers is still rather low. Most of them have problems in technically specifying the equipment required and in calculating the life-cycle cost of the investment. Offers are thus revised frequently, sometimes as often as five times. The largest part of the time with customers is spent specifying their needs. Since this is coordinated with the company in Singapore which makes the final design of the equipment, selling becomes very complicated.

The decision process involved in purchasing differs very much from customer to customer and is hard to diagnose. It is difficult, for example, to predict how many decision levels there will be and how much various individuals know about technical aspects of the products, how much they will bargain, and so forth. This requires thorough knowledge of the customer, which sometimes can only be achieved through close personal contacts.

Even if customers are gradually becoming more quality conscious, it has not been particularly easy thus far for the company to sell on the basis of quality, although this varies between different customer groups. European TNCs with Western purchasers, for instance, are indeed quality conscious. However, Chinese salesmen from Food Equipment often find this behavior very strange and it is difficult for them to handle.

The strong emphasis on price has cultural connotations. Chinese purchasers seem to act differently in bargaining with Westerners than they do when bargaining with other Chinese. Such basic conceptions as what is a reasonable price are completely different.

Social Linkage

Because of strong cultural barriers, only Chinese salesmen can handle relations with Chinese purchasers. Foreign managers or other expatriates do not come further than exchanging business cards with them. Business is done in the Chinese language. The main danger with local salesmen is that they become too much affiliated with their customers, making it difficult for them to change their priorities. Therefore, such a balancing of rapport and efficiency is considered to be the most important task for sales management in these countries.

Conmine's experience in how social linkages work in business relations illustrates a common experience of the TNCs which were studied (Jansson 1987). Similar patterns have been found in other studies.[1] For Conmine, business is more a result of social relations than vice versa, as it is in Western Europe. This has been illustrated at length earlier in the chapter. For Conmine, it is very difficult to get direct access to new customers in the engineering industry. Many visits are required for it to be allowed to make an offer. Even more visits are required to get an order. The social relations take anywhere from a month to over a year to develop. The length of this period is mainly influenced by the value of the potential sale and the intricacy of the buying process, or how decision levels have to be penetrated. It is also influenced by how strong an association the buyer has with competitors. The stronger it is, the greater is the time which is needed. It is also essential to mix with suppliers who service the same market, to obtain information from them. Broad contacts in many directions provide knowledge which makes it possible to plan better and to negotiate with different customers.

The main factor determining whether or not a salesman will get an order is rapport. This is a very personal trait. Certain social factors such as if the purchaser is a regular golfer or a schoolmate often play a role. This is an expression of "guanxi." However, the most important factor behind success in selling has to do with the market culture of the different ethnic groups, in particular the Chinese, the Malays, and those of Indian origin, and the types of social relations this creates. The last two groups are still uncommon in the business community of the area. There are small cultural differences between various Chinese dialect groups. Friendship generally is partly culturally determined. This is especially true in Malaysia and Indonesia, where business tends to follow cultural lines. A consequence of this is that most of the salesmen are Chinese, since purchasing (as well as trade in general) is managed by the Chinese.[2] "Buying your heart" is an expression there which illuminates how social relations are developed. Dinners are important meeting places, whereas lunches are not. In visiting a prospective customer for the first time, one observes such personal qualities as life style, way of talking and dressing, and ornaments, as well as how the office looks. Through observing cultural patterns here, important social information is obtained.

Friendship develops within the framework of social relationships which often are kept alive even if the parties involved move to other companies. This culturally determined social friendship goes deeper than professional friendship. If individuals know each other very well they can disclose the prices of competitors. Since this is considered unethical, however, a relationship of trust is a precondition to it.

Because of the long decision processes with many persons involved, it is very important to maintain rapport with many individuals all the way up and down the ladder of an organization, not simply with formal and informal decision makers, but also with such persons as those operating machines. Salesmen thus become very customer-oriented. Heated arguments are not accepted in discussions. They will never be forgiven or forgotten. Chinese often have long memories, which makes it difficult to repair mistakes.

Even small mistakes can be fatal. This is a consequence of harmony (face). Face was found in the present study to be of importance generally, both internally within the TNCs in question, and externally in relations with other companies (*cf.* Jansson 1987).[3] Personal relations tend to be more restrained and harmonious than candid and conflict-oriented.

The cultural factor is thus important to TNCs in marketing in SEA and it is particularly important in selling to local Chinese firms.[4]

Kickbacks

Another typical characteristic of purchasing behavior in SEA is the demand for kickbacks. This system seems to be particularly prevalent in two of the ASEAN countries, to be not prevalent in two of the other countries, and to be practically nonexistent in still another country in the region. It is also very much of an individual matter. Some persons are impossible to do business with without giving them a kickback. It is also important how competitors act. If they give kickbacks, a company that does not finds itself at a competitive disadvantage. Since it is part of the whole game of competition, it is important to find out whether it works as some kind of general fee for admission to be paid in doing business or whether one should reward a specific person to influence the decision in favor of one's company. In the latter case, information concerning how much competitors pay and to whom is essential. Usually the system works in one way in private industry and in another way in government, and it is very important to know how to use the kickback system in both cases. There are also differences in accordance with the type of products bought and the decision processes involved in purchasing. Governments almost always buy on the basis of tenders, where the major point of influence involves the writing of the tender specification. In private industry, building a continuous relationship is particularly critical. This helps a kickback system to thrive. In one Southeast Asian country the marketing strategy requires special tactical finesse in that kickbacks are part of the system. There is a need to "add chicory to coffee to make it go further." In some industries, the long life span of equipment, for instance, cannot be used as an argument to critical decision makers such as plant managers, who only expect to have their jobs for a fairly brief period, tending to get it shortly before retirement. This provides a great opportunity to the person to supplement a pension through kickbacks. Such incentives are normally taken care of by the agents. Hence, in the country in question this peculiarity of the social system particularly favors hiring an agent.

The kickback system is a real minefield for a TNC. Since this is corruption, a step in the wrong direction could blow up the whole local business and have negative international repercussions on the company's reputation in general. Many TNCs thus avoid getting involved in such a system. They feel it is better to use the money for social activities inside and outside the firm instead of paying it directly to certain decision makers and persons of influence. Those who do not avoid it seem to end up in a kind of "catch 22" situation. The system is hard to control, particularly by MDs from the West who are normally in the area on a three-year contract. Trying to control the costs of these activities, for instance by interfering with how agents operate this kind of business, may be dangerous, since such involvement increases the chances of being exposed. Avoiding control of these matters, at the same time, may allow matters to get out of control through the actions of others, which can result in loss of money by the company, perhaps with nothing in return, since no kickback guarantees a contract or a sale.

CONCLUSIONS

In this chapter, strategic marketing profiles were divided into a broad and a narrow category on the basis of the number of customers involved. Industrial marketing behavior was found to vary in terms of four different types of strategic marketing profiles. Narrow product specialists make a more detailed segmentation of customers than do broad product specialists. This was illustrated in the Pacmat case, which also indicated the importance of considering culture as a segmentation variable. Cultural influence was found to be a major factor influencing the social linkage and to be a cornerstone in linkage strategies. The Chinese, who are strongly oriented toward such linkages and their importance, display traditional ways of behaving in the business world. There is a culturally determined type of social friendship ("guanxi") which goes deeper than professional friendship. Face is also important. In addition, there is a cultural bias in the predilection for bargaining and for viewing price as the main competitive variable. This bias can also be noted in buying behavior, particularly in shortsighted behavior. Purchasing also involves a rather low degree of technical knowledge on the part of those involved, as well as complicated decision processes. Another culturally determined aspect of purchasing which influences competition is the kickback system. Whether TNCs like it or not, they have to relate to it in one way or another.

The social linkages and contact nets which are established are characterized by a high degree of informality in business relations. This underlines the exceptional importance of personal selling as the primary medium of transacting. The establishment and maintenance of contact nets has been illustrated for several TNCs. The more products are specially adapted to the needs of the individual customer and the fewer customers there are, the broader and more intensive the contact nets and individual buyer/seller linkages tend to be. Such high social impactedness yields a high specificity of linkages. A major result of the study can be seen as that of indicating the critical importance of individual and social trust as compared with organizational trust and professional trust in industrial marketing behavior in SEA.

NOTES

1. This was demonstrated in Chapter 2, particularly in note 7.

2. Redding (1982, 110-11) analyzes the Chinese element in buying/selling situations. "It is likely that a Chinese manager would bring to a business transaction some combination of the following (in no particular order of importance).

1. Desire for wealth as a source of security.
2. A family consciousness which presents him with obligations which can best be fulfilled by maintaining company prosperity.
3. A sense of Chineseness which is always obtrusive but is nevertheless deep and influential. It can also come out as an anti-order group feeling.
4. A sense of dignity and sensitivity to 'face,' often emerging as concern over rank (i.e., what rank is the person sent to deal with him).
5. A desire to avoid conflict, to maintain social harmony.
6. A desire to establish a friendship style of relationship in a business transaction.

7. A pragmatic sense of getting things to work by seeing different situations in different ways, not resting on any universal set of guidelines.
8. The assumption that trust is important and to be used, and a consequent disdain for the contract type of relationship."

3. See also Redding (1980). Redding and Ng (1983) give a good account of how face works within Chinese companies in Hong Kong, an account which also seems to be largely valid in Southeast Asia.

4. A good summary of the main Chinese forms of organization is found in Redding (1980, 138):

1. Intuitive, contextual, immediate decision-making, without a formal planning framework.
2. Informality of organization structure.
3. Low objectivity of performance measurement.
4. Personalistic external linkages to suppliers/customers.
5. Nepotism, patronage and cliques internally.
6. Centralization of power.
7. A high degree of strategic adaptability. (Redding 1980, 138)

Chapter Six

Competitive Strategies

Price is the most important factor in competitive strategy in Southeast Asia, and TNCs operating in the area need to adapt to this market condition. A common strategy of the TNCs studied was to reduce the importance of price by emphasizing quality. This and other competitive strategies of the TNCs in these special Southeast Asian markets will be described in detail in this chapter. The first section is a direct continuation of the earlier analysis of the strategies of product specialists and customer specialists. This is complemented in the second section by an analysis of the competitive and linkage strategies of distribution specialists and distributor specialists.

THE MAIN COMPETITIVE FACTOR: PRICE

This study indicated high transaction costs for the marketing of technologically complex products. Costs are substantial for determining prices, for bargaining about them, and for enforcing the agreed prices. The price levels agreed upon here are mainly a consequence of the product's fixed quality level. Prices are generally fixed for long periods of time because of the high costs of recalculation. The role of price in the marketing strategy of the firm is constrained in the case of customer specialists and product specialists. However, the markets in SEA have a strong demand for inexpensive products, as price there in general is the prime competitive factor. The major marketing strategy of the TNCs studied then was to try to reduce the perceived importance of price by emphasizing factors related to the needs of the customer, and the complexity and sophistication of the products. These included various aspects of competitive strategy which concern the quality of the product, such as technical advance, after-sales service, the means for achieving high-quality performance, and technology transfer.

The transaction costs involved in marketing high-tech products are higher than those connected with simple standard products, as the market situation for products of the latter type is reminiscent of market governance in that price is the dominant information carrier. The greater the importance of price, the lower the level of transaction costs. The higher uncertainty in bilateral and trilateral governance generally makes transaction costs there higher. This is the main problem for the TNCs studied here, that is, how best to reduce the high transaction costs that a competitive strategy requires. The relation between price level, which is mainly a consequence of quality level and of the nature of long-term relationships between business units, on the one hand, and transaction costs, on the other, varies for different types of needs. Transaction costs are higher for special customized products than for standard products. A high degree of linkage specificity complicates price setting. The strategic situation for customer specialists and for product specialists differs here.

Price and Complexity of Needs

Price is coupled with various factors in the model such as type and complexity of need, type of product, and the presence of a high degree of linkage specificity through long-term relationships. To satisfy a particular need a certain performance is demanded. Costs for discovering a need and acting in accordance with the discovery vary with the complexity of the need in question. A need can be satisfied gradually, starting with one function and adding others later. In the beginning, the buyer is interested in satisfying a core need through purchasing the product. Later, auxiliary needs come to the fore. The buyer either discovers them alone or is helped to realize them by the seller. Service and maintenance are examples of such needs. The complexity of needs may also change over time. After having experienced how a certain product fulfills a need, a buyer may demand either a more complicated or a less complicated product the next time. The needs of a customer may also change with industrial development in the country. The seller can adapt to this by offering various packages, where different features are added as the demand for them evolves.

The economics of transacting are dependent upon the complexity of the need. If buyers in a developing country, for instance, are not aware of a certain need and how it can be fulfilled through a sophisticated product, the transaction costs of creating it may be prohibitive. If this is the case, their uncertainty involved will be too great. This can be summarized in the form of a strategy matrix pertaining to strategic situations found within bilateral and trilateral governance. Such a matrix assumes that the product is a major result of the relationship as well as the main solution to the customer's needs. The product complexity this requires is often high. For example, there is an association between degree of product complexity, the presence of a long-term relationship, and the utilization of quality ahead of price as the primary component of the competitive strategy. Information costs, bargaining costs and enforcement costs are usually high due to the complexity of needs. Such costs come about chiefly through efforts to lay down the conditions for the product and its transfer. Price is not normally fixed before this has occurred. It is a consequence of the quality demanded by the customer and other product-related requirements such as service. When prices reflect these transaction costs they are competitive. However, because of the high uncertainty normally present in the bilateral and trilateral forms of governance, the matter of whether a price is cost-efficient is hard for the parties to control. If cost-efficiency is to be achieved, the parties must know each other very well. The more heterogeneous a product is, the higher the linkage specificity is and the more difficult it is to check the reasonableness of the price, since comparable alternatives are missing. For the buyer, this can be expressed as a high cost of determining the price of the product. The cost can be reduced if the buyer has close contacts with the seller and so can make certain that the price is reasonable. It is also important that the buyer get to know the behavior of the seller and whether the latter can be trusted. There is thus a mutual interest in controlling each other in order to safeguard business. How much to invest in a given relationship is contingent upon how it is related to other relationships, that is, its degree of linkage specificity. The higher the asset-specificity, the lower the degree of control by the market and the greater the investment in a specific relationship. However, since not everything can be known about the other party, there is a limit to how much is economical to invest in trust-building activities. Too trustful a relationship may even be inefficient, partly because this could mean that individual buyers or sellers identify themselves too much with the other party instead of with their own company.

In bilateral exchange, in particular, it is expensive to change prices, and prices will thus only seldom be altered. It is costly to find out about changes in the other party or in market conditions which could motivate a recalculation of price. The administrative costs of this are also considerable, particularly in a large TNC in which many units are affected. Furthermore, it is hard to motivate a price change to the other party. An arduously established trust could be destroyed by unmotivated price changes. An established relationship is characterized by habitual behavior, partly manifesting itself as a price norm that is hard to change. This results in stable relationships and market-type situations. The impact of price changes emanating from the environment varies among other things with the complexity of the relationship. Often a certain procedure is followed, even here. Prices are normally fixed for a year. Competition is usually segmented according to how expensive it is to detect the prices of competitors. One common way to test the market is to request offers from competing sellers in order to check their prices.

COMPETITION

Due to specific competition and buying behavior, companies seem to encounter a different market situation in SEA than in Europe. Competition is usually stiffer. Several TNCs from all over the world often become established in any given geographical and product area, representing perhaps the major transnational corporations in the product area. There are also competitive local firms in the market, run mainly by the Chinese. Japanese competition is normally strong too, although this varies with the industry. In some industries there is also a strong rivalry from Taiwanese and Korean companies. Competition is fierce in SEA, usually more intense than in Europe. For Pacmat, a narrow product specialist, on the other hand, the opposite is true. Indeed, there is no rule without an exception. That company, which has been active in the region for many years, has a large first-mover advantage involving an estimated lead of some ten years in both technology and size. For the companies studied the after-market is usually important too. Pirates which sell copied spares are very tough competitors for a few of the companies, causing them much trouble.

Another general observation from the study is that Asian competitors behave differently from their Western counterparts. Their propensity for price competition is greater, for example. This makes them more responsive to local demand. Factors such as the lower competence of most buyers in evaluating technically complex products and a different business culture result in different buying behavior than companies are familiar with in Europe. This dissimilar competition and buying behavior favor a short-term approach to business with not very technically advanced products at low prices. The complexity of the products which the companies studied tried to sell through changing the habits of their customers, that is through increasing the technical competence level and improving the buying routines of their customers, was high. The way in which companies adapt their marketing strategies to such a situation is described below.

Competitive Strategies

A major issue of competitive strategy for the TNCs studied was how to balance price against quality. Continuity in the shape of the price pattern and low and stable prices is a typical norm in most markets. Economic conditions clearly affect competitive strategy. It has also been shown that the proper strategy can change with the business cycle. In Southeast Asian markets demand changes very rapidly. The downturn at the end of 1984 and the beginning of 1985, after a long and escalating boom, was very sharp and extremely unexpected. The upturn beginning in 1987 was also quite rapid. One explanation of this pattern is that industrial markets in the region are still rather small.

Other major factors of influence are customer needs and customer buying behavior, the latter evident, for example, in purchasing routines and how these are related to the level of technical competence. Indirectly, purchasing behavior is also influenced by the economic conditions. In some of the companies, the competitive strategy is adapted to differences between various groups of customers. It is generally difficult to sell on the basis of the major arguments employed in Europe or North America regarding the costs customers can save through the high efficiency and long life of the product. One solution to this problem is to upgrade the customer through the transfer of know-how or through investment in service. Easy accessibility to the customer through geographical location is also critical, allowing for the establishment of a network of close social contacts. As illustrated in examples below, reliability in the sense of trust compensates for a lack of information and opportunism. The major option involved in this customer development strategy is to adapt products to the lower technical level of the customers, chiefly to achieve more flexibility in price setting.

Price is the most important component of competitive strategy in SEA. This was particularly true during the period of deep slump from 1985 to 1987. During those years it was especially difficult for the TNCs studied to sell their technically advanced high-priced products. As a result, many companies ran into serious difficulties. Those established in boom years had problems adapting to a completely different economic situation. Some even thought of closing down their subsidiaries. Competitive strategy is not favored by low demand caused among other things by a lack of money. It does not seem to be favored either by booms. In the earlier boom, which lasted until the end of 1984, rapidly changing demand and a permanent shortage of products were common, providing excellent opportunities to earn quick profits. Interest rates were also high, serving as an impediment to investing in long-life equipment and in products utilized by such equipment.

Thus far, the economic development occurring in SEA, during both slumps and booms, seems to have fostered short-term business behavior. This is in line with the business culture of the Chinese, which has its roots in the commodity trade traditional in SEA. In Less Industrialized Countries the development of an industry tends to be coupled with a very uneven, and basically low technical level of production equipment. The same is also true of the knowledge, experience, and patterns of thought of many people in the customer organizations. Buyers, therefore, tend to lack competence in analyzing matters of production cost and efficiency in relation to various practical requirements. This is often made apparent by an inability to calculate the profitability of an investment. There is also a lack of demand for after-sales service. Customers are not interested in maintaining their equipment. These circumstances make it costly for sellers to use the typical competitive tools of a high-quality producer: training (transfer of know-how) and service. In the beginning, due to these factors, there are definite limitations on the sales which can be made to those segments of the industrial market in which a demand for sophisticated industrial products exists and which have the innate capacity to make use of such products. Other TNCs in the area are a customer group to which it may be best, at that point, to largely turn. Then the market base can be gradually broadened to increasingly include less-qualified customer segments.

The ability to compete is strongly affected by how marketing activities are organized. Most of the TNCs studied had set up their own subsidiaries through internalizing marketing. The high-technology profile of these companies, together with a competitive strategy built on nearness to customers and close relationships with them, requires both technically qualified salespeople and technical experts who are based in the region. This is most efficient to organize internally. Since the break-even level for internal organization is higher than for external organization, it is also essential to have a broad product program. This was evident in a comparison of the companies studied with many of their competitors. In establishing an internal organization, a basic aim of these companies was to create a market platform capable of generating a profitable and sustainable business. Through the organization this requires, TNCs commit themselves to operations in the host country on the basis of high investment. In this way, the competitive strength needed to sell advanced products is increased at the same time as the potential for setting low prices is unfortunately reduced. Competitors, on the other hand, through their more flexible external organization, have an opposite profile, providing them better opportunities for price competition. Such a difference was also found between the European TNCs studied and Japanese companies. The European TNCs were considered by many of their customers to be superior in the transfer of technology.

Service is an important but difficult aspect of competitive strategy. The transfer of technology through increasing the technical level of customers is an equally important condition for the sort of competitive strategy most of the TNCs studied pursued. The importance of bridging the temporal gap between the seller and the buyer through the coordination of production varies a lot from company to company. Logistic matters such as delivery time and transport are essential for broad product specialists. Such matters are even more important for distributor specialists, for which availability of the products with intermediaries is critical. Such companies usually have a warehouse centrally located in Singapore from which products are distributed throughout the region. For one of the TNCs, having production located close to the market is an important competitive strength, just as basing technical service and marketing on a close knowledge of local conditions, providing training of customer personnel, and maintaining good financing conditions are additional competitive strengths.

Financing was only found to be important for narrow customer specialists selling to large projects and then operating within the framework of trilateral governance. The marketing support provided them by Export Councils and other public organizations was of no appreciable consequence. The large TNCs studied are normally self-sufficient and do not require much help of this type from outside parties.

A conclusion to be drawn from the study is that an institutional factor such as infrastructure does not create any problems of note for business operations in Singapore and Malaysia. In Indonesia, the Philippines, and lately in Thailand, the situation is different. Deficient infrastructures there create marketing problems, but also opportunities. An inexpensive distribution network there could provide a distribution specialist with a competitive advantage. Bad communications serve as barriers for competitors not established in the market. Under such circumstances, decentralization tends to be a key element in marketing organization. Locally based salesmen are trained to be more or less self-sufficient.

As already indicated, price and other competitive factors are related to linkage specificity. The higher the linkage specificity, the greater the increase in switching costs and the more isolated the linkage. Linkage specificity is affected by other factors such as the type and complexity of the need and the product type. Strategic profiles are closely related to linkage specificity and have a strong influence on competitive strategy. In the examples to be discussed, one broad customer specialist (Food Equipment) will be compared with two broad product specialists (Conmine and Tooltec). As already suggested, however, there are a number of important differences between various of the firms. These are illustrated in the case of two broad product specialists. These differ both in sales concentration and in the technological sophistication of their products. Further examples are drawn from these three primary cases.

BROAD CUSTOMER/PRODUCT SPECIALIST

The competitive strategy of Food Equipment stresses quality. However, compared with the other two primary cases considered below, the concept of quality is broadened in this company's case to include the whole life-cycle of products, thereby focusing on the need instead of on the product as such. Food Equipment is more likely to adapt its products to local needs than many of the other TNCs studied. It is a broad customer specialist, but also a product specialist. It sells minor equipment and certain major equipment, adapted somewhat to individual customer needs, to raw material and food processing industries. Some of its major equipment is sold to large projects. The after-market, products such as spares and service, are also of special interest to the company. Like many other companies, Food Equipment has experienced how quickly markets change in SEA and has learned how to adapt to this situation. For equipment sold to the raw material processing industry, changes in demand have been extreme with large and rapid demand swings between marked booms and tremendous slumps.

In general, the company's competition is product- or systems-oriented. However, the competitors' market coverage is normally not as great as Food Equipment's. The company is faced with basically the same competitors in SEA as in the rest of the world. In addition there are Japanese firms, companies from the Peoples Republic of

China, and local Chinese companies on the market. Japanese corporations are the most competitive, since they sell good products at low prices. A common Japanese strategy, as in one Thai market, for instance, is to sell equipment at a low price and to later compensate for that by employing high prices in the after-market. This also creates continuous contacts with customers through service people, something which is valuable in future business. The unwillingness of Japanese corporations to transfer technology to customers and their inflexibility in finding new solutions to customer needs outside of their standard solutions are weak points.

Companies that sell products at low prices and that are adapted to the right quality-level get a decisive first-mover advantage. Price is a very important factor in the competitive mix, especially during slumps such as those in 1985 and 1986. For the high-quality segment, price is of no great importance. This segment has grown in recent years with the rising export dependence of such customers. When products are exported, their quality level must be up to international standards. The basically most satisfactory combination of price and quality is expressed by the term "reliability," which is a broad concept covering spares and service as well. It represents high quality generally, and adaptation to the technical level of the customer's factory and the skills and knowledge of the workers there. In that way, the product and the customer are adapted to each other, reducing the need for maintenance. Quality is developed through nearness of the company to the customer. It is not possible to obtain any results from an increased technical presence, however, before social contacts have been established. After that, customers develop their technical and purchasing capacity. Once trust has been established, customers know that they can come to Food Equipment and discuss their problems. First-mover advantages here are mainly obtained by investing in customer relationships. This, in turn, contributes to the development of reliable products.

To make this possible, an engineering capacity has gradually been generated centrally in the region, making it possible among other things to prepare offers for the local market. In one product area, technical capacity originally based in Europe has been transferred to SEA for the development of major equipment. The marketing of such customized equipment takes place in cooperation between the local market companies, which maintain customer contacts, and the regional unit in Singapore, where the engineering capacity is located. For products for which very little engineering expertise is needed, the market companies manage on their own. For equipment, for which much engineering expertise from outside the region is required, marketing is coordinated between units in the region and in Europe. The question of how to best organize the connections between the different units of a group in matters concerning R&D, product development, and customer needs within the international market is one of the most crucial issues for the survival of a high-technology TNC such as Food Equipment.

The After-Market

The after-market is important for a TNC since the selling of original equipment generates a market for service and spares. In Malaysia, this market has involved the sale of large amounts of equipment over the years, as the wear and tear on equipment is considerable. This large after-market volume for rather uniform equipment in a small country has created special problems for Food Equipment. Large numbers of small local firms were established for service, repair, and spares. Some of these entrepreneurs were previously employed by Food Equipment. They left the company and took their knowledge with them, using it to copy Food Equipment's products and services. These pirates are very competitive. Food Equipment's market share in Malaysia is only about 35 to 40 percent as compared with a normal 90 to 95 percent in other markets. The success of the pirates is partly due to encouragement they have received from customers, who apparently hope to increase competition in the market. The problems are aggravated by the fact that customers are not very quality conscious.

These competitors operate in a different way than does Food Equipment. Their organizations are very small, the manufacturing equipment is old and requires little overhead, and labor costs are low as well. The firms in question concentrate on the most profitable parts of the market. These are small Chinese family firms, normally with no more than 15 employees which often sell to other Chinese firms. They speak the same language and have a common culture. These industrious entrepreneurs are proficient in copying everything with a high accuracy for detail, without using any blueprints. To some extent this seems to be part of the culture of the region. Pirates of this sort are now spreading to other Southeast Asian markets, where the original equipment has not been available so long. The family bonds of these entrepreneurs facilitate such an internationalization of business activities.

Broad Product Specialists

Conmine sells standard equipment of major and minor character, together with MRO items. These products are imported from Europe with no attempt to adapt them to markets in SEA. One reason for this is that markets there develop very quickly, and so products demanded today may not be demanded tomorrow. In principle, the full range of the group's products is open for sale, but in practice a limited range is sold.

The main competitors in the Singaporean manufacturing industry, together with competing TNCs from the U.S. and Japan, have a market share of about 50 percent altogether. Other, mostly local, firms have a market share of about 20 percent, and Conmine has the rest. Regarding the major equipment, for example, U.S. firms are the strongest on the advanced, high-technological side. Japanese firms have improved their position after having been most competitive at the lower end of the market.

In the construction industry, where mostly standard products such as minor equipment and MRO items are sold, the Japanese competition is the strongest. The major competitor, which is American, is active here. Just as Conmine does, the company markets a full range of products, whereas the other companies compete with more limited product ranges. Although the competitive situation in Malaysia is similar, the major part of Conmine's sales there is to the manufacturing industry. Conmine and its main competitor both have subsidiaries, whereas the other companies are represented simply through distributors. Japanese competitors are generally stronger in the Far East than in other parts of the world.

As viewed by Conmine's customers in the construction industry, the major components of competitive strategy, in order of importance, are: price, times of delivery, finance (terms of payment are important here), quality, service, and lastly training. This ranking is especially relevant to Japanese customers and to local Chinese firms, to which it is particularly difficult for Conmine to sell in the traditional way, on the basis of quality. Even if price is the main aspect considered by large international contractors, they are more receptive to arguments of quality than are the Japanese and local Chinese customers. These contractors usually have well-organized purchasing organizations that make strict evaluations of equipment which is offered. Equipment is bought and utilized for specific construction projects in the area. The limited life span of the projects results in shorter pay-off periods than in the manufacturing industry. Equipment is depreciated in a shorter period, which makes price important. Because of this shorter period it is easier to make cost calculations and to relate them to the immediate price of the product. In the manufacturing industry, the ranking by Conmine's customers of the elements of the competitive strategy are somewhat different: price, quality, service, times of delivery, finance, and then training.

Conmine's competitive strategy toward customers in both the construction and the manufacturing industries emphasizes product quality and various matters associated with it, such as service and training. This strategy is used to defend the higher purchasing prices of their products as compared with their competitors. Their selling argument is that customers are compensated through the longer life of the product and lower operating costs, as the whole life-cycle costs are lower.

This strategy succeeds readily enough in Singapore. In the Malaysian manufacturing industry, price is so important that it is difficult for Conmine to be successful using its traditional strategy. There, price is especially important when the first piece of equipment is sold. Sometimes the company sells at a very low price in order to get a first order from a customer and thus start the relationship. However, the importance of this has gradually decreased with the growing technical capabilities of the engineering industry, particularly as increasingly advanced industries are established in the country. In 1989 the situation improved as compared with 1985. In the mid-1980s it was rather difficult to compete except by selling to certain types of customers. Primarily it was TNCs with basically the same purchasing policies in Malaysia as in other countries which bought the company's high-quality products at its high prices. Most customers did not have the ability to differentiate between various products and to assess their differing qualities. They perceived only more crude differences in quality between the products of various companies, for example between European and U.S. TNCs, on the one hand, and Japanese companies on the other. More customers from Europe and the U.S. than from Japan were willing to pay the high prices for their products. Therefore, Conmine concentrated its competitive strategy on TNCs within that first group.

The more demand declined during the recession in the mid-1980s, the more difficult it became to sell on the basis of quality. The previous price difference was not so large that it prevented many customers from paying for the higher quality involved. With increasing price competition and declining prices, however, the price difference more than doubled in a short time. This made it difficult for the company to sell its high-quality products, even to its oldest customers. The time spent on price negotiations increased, since more and more customers came back for better prices after having negotiated with competitors. This caused the marketing strategy to be quickly changed, since demand decreased very fast, more or less overnight. The main reason for this was that the industrial base in Malaysia at the time was quite small. Due to the larger industrial base in Singapore, changes in the market there were less rapid.

After-sales service, the second most important component of the competitive strategy of the manufacturing industry, is difficult to use as an incentive in most Southeast Asian markets. It cannot be taken for granted that the offer will be as attractive as it is in Europe. The service potential is large, but the actual need experienced is low. In general, customers are not willing to pay for it. Despite this, Conmine has invested in a service program in one of the main SEA markets, since a higher price must be justified by extra advantages. However, the market was only ready for this five years later. This program has made it easier, nevertheless, to base competitive strategy on continuity and reliability. At the same time, a competitive advantage has also been created that is hard to imitate. The main problem in pricing service is the low labor costs in LICs. A maintenance contract, therefore, tends never to specify this cost. Thus, the company has managed to build the service component into its competitive strategy for the manufacturing industry, where interest in continuous supply and in quality products is larger than in the construction industry. In another of the countries, where the more advanced sector of the manufacturing industry has expanded a great deal in recent years, it has become easier recently to elicit support on the basis of the service concept. More and more service contracts are being sold there and they are even often priced separately following negotiations.

Service is a broad concept for Conmine and represents much more than simply an after-market. In its broadest sense, service involves finding applications that justify the equipment sold. It includes assistance of all kinds given to customers to reduce the life-cycle costs of an investment. The customer is offered a free start-up, a priority-access to a service center (if a service contract is part of the deal), technical backup when buying equipment, and priority-access to spare parts. Training both before and after sales is also part of the service. It is important to train the customer to purchase high-quality products. Often, the tendering process itself becomes a training course.

Another broad product specialist, Tooltec, follows a strategy similar to Conmine's. The major products, technically advanced minor equipment and processed materials, are marketed to the engineering industry as high-quality products. The main argument for justifying higher prices is the lower total cost which results for the customer's manufacturing. Purchasing prices for these products are around 30 to 40 percent higher than those of competitors or than the prices of simpler versions of their own products. Therefore, the potential for cost savings for customers must be high. The basic problem here is the same as for Conmine: to find customers with an interest in such products despite their operating in an environment in which most products are viewed as commodities. Thus far, Tooltec has sold mainly to medium-sized customers and to some large customers spread unevenly over SEA. There are large differences between the different markets in the region, as sales in most of them still are rather low. Singapore is an exception, since the engineering industry there is more highly developed. Within a given market there are large differences in the quality of the products available. Markets can be divided into several layers or segments, from the most advanced segment, where Tooltec is active, to particularly low-quality segments.

Tooltec's products are imported from Europe. They are the result of applications based on customer needs in the major markets of the Industrialized Countries. It is important that the company determine to what extent these applications are relevant in Southeast Asia, and whether they are competitive and profitable enough to justify efforts to create a market for them. Only a limited range of the group's products can be sold in the area. Over the years, experiments have been made in selling price-competitive lower-technology products. The results have been mixed. This is also contrary to Tooltec's main image as a high-technology firm selling high-quality products.

Competition for the company is very stiff. For minor equipment, in particular, there are many competitors, such as Japanese firms and other European TNCs. For processed materials, Japanese companies are by far the strongest competitors. The region can be seen as a Japanese home market.

In its sale of processed materials, Tooltec has found it easier to compete with other European companies than with Japanese firms. This is particularly true in the case of large projects, which involve a strong concentration of sales to a few customers. Japanese firms, backed as they are by the state (see Chapter Seven), are very strong here. Price continuity is very important in the sale of processed materials. Business is built on fixed prices and on long-term relationships between sellers and buyers. If prices were raised, trust could easily be lost. The strategy of price continuity, to be sure, is followed by the Japanese. Such a strategy would be difficult, however, for a sales company that imports its products from Europe to manage. Many product companies believe, nevertheless, that for the marginal quantities sold in Asia it is possible to employ the same price policy as in Europe, that of reducing prices during slumps and raising prices during booms. They also frequently seem to assume that markets in Asia are in the same phase of the business cycle as the European market. Behavior based on such notions is not accepted at all in the region. Relations that took many years to create could be destroyed overnight if an already-established company behaved in this way.

Price considerations have the highest priority for all of Tooltec's products. Next come quality and availability (times of delivery). The latter competitive strategy component is particularly important for minor equipment which is technically less advanced, and for MRO-items sold to distributors. For the minor equipment which is technologically more advanced, equipment which is sold directly to the engineering industry, service is also very important. Customers are normally interested in getting service but are unwilling to pay for it. Since the company does not want to give service away free of charge, service is included in product prices. The training of customers is thus critical for competitive strategy here. However, it is an arduous task to get customers to adopt modern manufacturing methods. There is a strong resistance to abandoning old and established habits. Furthermore, both labor costs and evaluation of the time factor are different in SEA than in Europe and North America, a situation which fosters low efficiency requirements. Since 1985, it has become slightly easier in certain markets to sell on the basis of quality. There is also more interest in service and in high-quality spares than previously. Purchasing engineers have become more and more experienced. TNCs with purchasing strategies which are basically the same throughout the world have also invested heavily in the region. However, there is often a lag in business opportunity, since the company's minor equipment is often sold in the after-market and not as original equipment. The company has found questions of financing to be important when selling to large projects. The terms of payment usually allow credit of more than 100 days.

Another major lesson Tooltec has learned has been the importance of developing the image of the company and of having a high-quality profile. Although internationally this is one of the best-known TNCs in its areas of business, it was completely unknown in SEA when it started business there. The company's reputation had to be established for each of its major product groups.

DISTRIBUTION AND DISTRIBUTOR SPECIALISTS

Distribution specialists are similar in their strategic profiles to the customer and product specialists discussed above in that they have direct contacts with their customers. All three types of specialists have integrated forward to the customer, so to speak. An important difference concerns the importance of matters of distribution. Distributor specialists, in contrast, tend mainly to have indirect contacts with customers. Among the TNCs studied there are only two pure cases of these two types of specialists on distribution: Specmat, which is classified as a distribution specialist, and Indcomp, a distributor specialist. Another company, Tooltec, has distribution as one of its main strategic marketing profiles and the minor marketing activities of a few of the other companies characterized them as distributor specialists.

A Distribution Specialist

Only one of the TNCs studied is represented in Asia (and in Australia) by a distribution network of its own: Specmat. It consists of eight sales units established in the major Asian markets. These units are controlled from headquarters located in Singapore, which were originally located in Europe, and all sales to markets outside Europe and North America were controlled from there. With the growing importance of Asian markets, however, nearly the entire headquarters for this region was transferred to Asia and was integrated with the sales companies there into an independent operation. Specmat has sold the processed material in question in Asia ever since the 1940s. This work, first performed by agents, was gradually taken over by Specmat's own sales companies. The first such company was established in Japan in the 1950s; the second (Singaporean) company was founded in the early 1970s; and the other sales companies were established during the 1980s.

Specmat originally represented other European TNCs as a distributor. Subsequent restructuring of European industry resulted in only one supplier and owner of the company. This situation recently changed when this owner merged with a large European TNC. This will surely change Specmat's Asian operations, for new products will undoubtedly be added to existing ones.

Specmat's main marketing functions are the distribution and sale of a special processed material to a wide variety of customers. Some 20 different types of the material are sold, each of different quality, with approximately 100 dimensions to each quality. The average order is extremely small for this type of product. The service level is high, as customers frequently demand delivery within 24 hours. The material is shipped in bulk from Europe and is transformed on the spot in accordance with customer needs. Due to the great diversity of the high-quality material involved, sales activities are particularly important. Traveling sales representatives and service engineers, together with salespeople at the various offices, form a contact net with customers. Those traveling and primarily maintaining contact with customers amount to about 20 percent of the company's employees in Asia, whereas about 30 percent of the employees there are engaged in matters of distribution. The rest of the employees are mainly office salespeople, managers, and clerks. Contact nets are essential. The software aspects of the overall product and the emphasis on quality are used to justify prices clearly above those of competition. Another factor necessitating high prices is the high training costs for sales personnel. Such costs, unfortunately, could increase even more if these salespersons were hired by competitors, or if they left to open a competing agency. This risk is much higher than it would be in Europe. Because its marketing strategy is executed by a sales organization of its own, it is possible for the company not only to keep prices high but also to maintain a long-term orientation with rather steady price levels. This is very important in an industry in which marginal pricing and even dumping are common, and under conditions of a business cycle which is quite different from that in Europe. Pricing and dumping practices are often put into effect with the help of agents. A company which sells through its own sales organization, in contrast, must have a much stronger interest in the long-term business aspects.

On the basis of its large marketing organization in Asia, Specmat possesses a powerful potential for coordinating sales in the region. This is of increasing importance as regional integration of the customer industries there grow. Such integration generally occurs to a considerable degree through the emigration of industries from Japan to North and Southeast Asia. Japanese industry is becoming dominant in East Asia; Specmat, in particular, has transferred many sales people from Japan to Southeast Asia to sell to Japanese companies there.

Distributor Specialists

Industrial Distributors

For Tooltec price, availability, and service are the most important components of its competitive strategy regarding distributors. Training is also important, especially to build up the Chinese distributors' overall knowledge of marketing. Such training is adapted to the specific business culture of the Chinese. This demands long-term relationships that often are not easy to establish. The loyalty of distributors (also referred to here as intermediaries) is low. Business with them can therefore consist largely of individual deals which are renegotiated every time.

Tooltec has little control over intermediaries' pricing to final customers. Intermediaries set prices on their own without informing the seller how costs are estimated. However, certain insight into this can be obtained in negotiations, particularly when intermediaries demand lower prices, which they can be forced to justify. On the whole, however, subsidiaries are responsible for conducting their own investigations of markets, mainly by checking on final customer prices.

Tooltec's price negotiations with intermediaries are often very arduous. In contrast with direct negotiations with final customers, little comparison of prices with competitors is possible. Rather, the distributor's intimate knowledge of the market is the starting point for much of the bargaining. The pricing adopted by these intermediaries is partly controlled through a discount system. Tooltec designs such a system to prevent distributors from buying more products than they can sell. In that way, competition among intermediaries is avoided and at the same time there is a continuous flow of business. Discounts tend to be differentiated according to the size and the type of distributor involved. Tooltec tries to provide other types of support as well to increase the flexibility of service levels. It plans to learn more about its final customers and their capacity directly rather than continuing the traditional focus on separate orders and products.

Industrial Distributors and Own Direct Sales

One very large TNC, Indcomp, is the market leader in its field of business in Western Europe and North America; However, it is far from that in Southeast and Eastern Asia, the domain of the Japanese companies. Competition has made it particularly costly for Indcomp to try to gain access to the high volumes inherent in direct selling to equipment manufactures. It is very hard to enter preexisting relationships between Japanese manufacturers and suppliers in both Japan and SEA. The company's initial strategy has thus been to concentrate on the low-volume industrial after-market, selling mainly through intermediaries. Indcomp offers several thousand different products or product versions, and is represented on the local level by various external parties, primarily distributors.

The regional headquarters and logistics center of the company is located in Singapore. The local activities of distributors for the larger markets are partly controlled through the company's own sales subsidiaries, which each has its own salespeople, product stocks, and administration. This provides the company with a better knowledge of the local market, improved possibilities for implementing its own policies with distributors, and better opportunities for supporting distributors and for visiting final customers directly than if sales had been centrally located. The activities of distributors for smaller markets, in turn, are controlled by a separate distributor company located in Singapore. When a given market has been developed sufficiently to allow Indcomp to establish its own sales subsidiary, the distributors in the area are transferred there. The new subsidiary is then made subordinate to regional headquarters, and partnership contracts with distributors are drawn. The degree of exclusiveness of a distributorship varies, determining the degree of control. The basic principle, however, is that the products Indcomp offers should be the most competitive and therefore those which are most attractive to distributors. The more attractive a product, the greater the extent to which hierarchical control can be replaced by market control, that is, by positive controls such as incentives, which are more effective than negative controls.

A very important component of Indcomp's competitive strategy is price, particularly in its competition with Japanese companies and when it sells to smaller customers through intermediaries. The strong competitive influence of the Japanese on prices is a major factor here. The emphasis on price is also strengthened by the fact that customers are often willing to waive quality. They perceive quality as relative, partly due to their orientation toward short-sightedness. If the time-horizon of a customer is five years, he is not willing to pay for a product that lasts twenty years. A problem for Indcomp in this market is that the share of Japanese companies in this product area is high and is also increasing since Japanese investments in SEA are on the rise. Establishment of the after-market business is also difficult, since many of the customer firms which have bought their original equipment from Japanese companies buy components as well from other Japanese suppliers.

The close location of business activities to customers is another important component of Indcomp's competitive strategy. Indcomp also stresses high quality as a major competitive tool for motivating higher prices. This is made possible by a strong local presence and a superior distribution network, for which proximity is clearly advantageous. The company's willingness to engage in technical discussions with customers and to train both distributors and customers are further matters which can justify higher prices. The ready availability of products in the region through a broad distribution network which enables the company to keep promised times of delivery is more important in SEA than in Europe. The central warehouse in Singapore can be seen as essential to the efficiency of the regional logistics system. The importance attached to this matter can be seen in the fact that Indcomp's logistics division is organized as a separate company.

Due to the intense price competition in Asia, Indcomp concentrates on its price strategy there much more than it does in Europe. Like most of the TNCs mentioned in this chapter, Indcomp began with a production-oriented cost-plus pricing which was adapted more and more to local market conditions. This is especially important in the industrial after-market, where price flexibility and transparency is high. The European price structure has been impossible to employ in this market, due to the different competitive situation in SEA. Although Indcomp's price levels are positioned above those of the competition, price differentials are much smaller than in Europe. Since distributors are independent price makers as regards the prices the final customers pay, Indcomp influences these prices through its own prices to distributors. This pricing has gradually improved from simple discount systems and is rather advanced today.

Pricing is also considerably more sophisticated in the after-market than in the original equipment market. The greater numbers of different types of products, customers, and industries result in a more segmented price strategy. Prices are controlled in one way or another in every stage. This has been achieved by a decentralization of the market organization and by the company's performing its own operations in the market. In that way a thorough market knowledge has been established within the organization. The market has been grouped into various segments according to price level and degree of price flexibility. Both prices to distributors and the type of marketing support provided are based on the final customer segments at which sales are directed. As already noted, Indcomp emphasize many different selling aspects besides price, such as product quality, availability (service level), maintenance, and after-sales service, in its competitive mix. A marketing program for distributors, developed by the company for worldwide use, has been adapted to the special market situations in SEA. The company's competitive profile is also enhanced through the upgrading of distributors and through bringing them closer to Indcomp. They are provided with many forms of assistance, and efforts have been made to establish a common identity and image for distributors. Through such a marketing policy, directed mainly at the high-price segment, it has been possible to raise prices on the average. Sometimes, however, for the low-price segments, prices to distributors are even below the full manufacturing costs. Because it is partners with them, Indcomp can affect the pricing of many of its distributors. The lower the price becomes, the more influence the company attempts to exert.

A major aim of Indcomp's marketing policy is to come closer to the final customers in order to provide the best possible solutions for their needs. This has become increasingly important, especially since Indcomp extended its marketing concept. Previously, the company focused on limited customer needs related to the selling and servicing of certain high-quality components. Now the focus is on solving customers' logistical problems regarding the transportation, storage, and production and process results of the end user. To fully attain this goal, Indcomp must extend its product range even more and concentrate still more on maintenance as a service. Close cooperation with final customers also requires a closer cooperation with distributors, such that the company and its distributors operate as a team. To this end, Indcomp is increasing its investments in the distribution network.

CONCLUSIONS

Competition in SEA is usually stiffer and more oriented toward price. The TNCs studied face three main types of competitors: other Western TNCs, local firms, and Japanese firms. Western TNCs primarily stress quality in competing, whereas local firms primarily stress price. An extreme example of the differences between these two emphases was presented in a description of the situation with which one of the TNCs was faced on the after-market. Difficulties with both strategies were illustrated in this chapter, for instance as regards service in the former case and product adaptation in the latter. The choice of strategy is influenced by the competitive situation. It is particularly hard for the TNCs which were studied to compete with Japanese companies which are strong in both quality and price. Thus, a major issue in planning the competitive strategy for sales of high-quality products is how to efficiently balance price and quality. One major option in SEA is then to emphasize quality at the expense of low prices. Another major option is to stress keeping the price down and adapt or downgrade quality accordingly.

The main connection between horizontal and vertical markets involves a fundamental transformation in which a large-numbers competitive situation is transformed into a small-numbers situation. This is analyzed here primarily in terms of linkage specificity. The higher the linkage specificity, the more narrow the horizontal market. The number of actual competitors may in this case be much lower than the number of potential competitors. This is typical of bilateral or trilateral governance. The strategic profile is one expression of linkage specificity and it can be used to relate linkage strategy and competitive strategy.

Difficulties in adapting products to Southeast Asian markets in a manner which allows the price to be lowered are greatest for broad product specialists and for distributor specialists. The main strength of a product specialist is the ability to market homogeneous quality in a broad assortment of standardized products. Marketing in SEA involves building relationships that create a certain linkage specificity. This was illustrated by the Conmine and Tooltec cases. For the customer specialist, this adaptation of quality and price comes more naturally, as in the Food Equipment case. However, it involves high investments in specific assets and a high information impactedness and social impactedness. Switching costs are also high, which reduces price flexibility. Since it becomes highly expensive to change to alternative suppliers, a greater reduction in prices is needed. This was observed in the Conmine case, in particular, under conditions of a worsening economic situation. For product specialists, linkage specificities are lower and price flexibility is higher. However, the situation of product specialists depends very much on how concentrated their sales are and how technologically sophisticated their products are. This was illustrated by the differences between the Conmine and the Tooltec cases. Price flexibility is highest for companies marketing technologically simple standard products to large numbers of customers. This comes close to market governance, in that price is the main information carrier and the major medium for resource allocation.

The marketing strategy of the sole distribution specialist is reminiscent of that of the broad customer specialist. The main difference is the stronger role of distribution, which, in the case of the former, is conducted jointly with sales. Besides price, availability and speed of delivery are important for distributor specialists. In the case of this strategic profile, linkages with final customers are replaced with linkages with distributors. Since this means only indirect contact with the market, the main problem is the distributors' control of the pricing. A similar problem is faced by product companies located in Europe when they deal with market companies located in SEA. The difficulty in controlling prices is partly due to the complexity of the products sold, which also strongly influences the importance of distribution as compared with sales. For simple products, where pricing is largely market controlled, the industrial distributor is faced with a market-like situation. Price control is achieved by a discount system. For technically more advanced products various types of support may be necessary as illustrated in the case of Indcomp. Long-term linkages need to be built with the distributors, allowing prices to be effectively controlled. The Indcomp case also demonstrates how contacts with different types of distributors can be segmented. Linkages with distributors are also instrumental in the company's direct contacts with end users. It is highly advantageous if a distributor specialist such as Indcomp, which combines distribution with joint sales contacts, has its own sales subsidiary located close to both distributors and customers.

This chapter is mainly a study of TNCs which sell technologically advanced products and the market investments of which are characterized by various kinds of linkage specificities. A location close to the customer, which creates strong links with the market, is a virtual necessity. A competitive profile consisting of high quality products, service, and technology transfer appears to require its own representation on the local market. A large part of the strategic profile tends to be located there and to be organized as a subsidiary, either concentrated on sales or combined with distribution. Even when there is an external organization, as with a distributor specialist, the seller is not far away and assists the distributor in sales. At the same time, TNCs with such competitive profiles are very much involved in price competition. The method of pricing products is a matter of critical importance to success in marketing.

Chapter Seven

Marketing of Projects

A considerable number of large infrastructural and industrial projects in Southeast Asia have created good business opportunities for many of the TNCs that were studied. Governments and other users have bought projects either as complete turn-key projects or as disembodied separate parts to be assembled by the buyer. During the economic slump between 1985 and 1987 sales of projects were poor, but they began to pick up again near the end of the 1980s when many projects which earlier had been shelved were implemented.

This chapter is devoted entirely to the marketing problems of narrow customer specialists. Most of the transactions analyzed took place in the trilateral governance form. In that way the theoretical framework in Jansson (1990) is developed further by relating it to the marketing economics approach. What characterizes the marketing logic in terms of this governance form? How do TNCs market their products and sets of projects? What characterizes the marketing situation and marketing process? What marketing strategies are employed? The marketing of a project is very much a question of the selling of an idea. A project does not exist in concrete form until it has been delivered and installed. The buyer must decide on the project and supplier, therefore, before determining how it will look when completed. During the greater part of the marketing process a project only exists on paper. The offer is a promise of service, times of delivery, price, etc., which can be first checked by the buyer only long after the placement of the order. This is an important fact, since it largely explains the high uncertainty experienced by both parties in this kind of business transaction. In idea-selling, the transfer of information and of social influence is more crucial in certain phases of the marketing process than the product itself. The buyer is certainly informed and hopefully persuaded of the superiority of the supplier's solution. However, it is the repeated social contacts which in particular reduce the buyer's uncertainty and increase trust in the seller. Good reputation, guarantees, and reference projects that can be visited by potential customers can have a similar effect.

The marketing of projects, as defined here, includes the marketing of all types of products related to a project; they can be individual products and services included in the project or a complete project itself, consisting of a package of products and services. Marketing is thus given a broader meaning than systems selling as defined by Hammarkvist, Håkansson, and Mattsson (1982, 92). A system is a package consisting of both hardware and software, separate from the sale of components and other individual products. The package deal is another closely related concept (Ghauri 1983, 16-17; Hadjikhani 1984, 5-6). Here the project can be bought packed or unpacked. Hence, many types of products–of components to projects as well as of different types of package deals–may be involved.

This chapter is divided into three main sections. First, various situations for the marketing of projects are described and illustrated. This is followed by a section on another specific aspect of the marketing projects, the linking process. The final section covers linkage strategies and competitive strategies. Selling to projects is an activity mainly undertaken by other TNCs than those used as illustrations in the earlier chapters. Four companies, Paving systems, Weldprod, Telecom, and Indpow, whose activities have been only marginally illustrated before, are taken up here. In addition some examples are taken from better-known companies such as Food Equipment and Tooltec.

MARKETING SITUATIONS

The chapter is limited to marketing situations in which the TNC which sells manages the business of the project itself. Consortia and similar cooperative arrangements between independent firms are excluded. Two main types of marketing situations are analyzed. One is similar to those marketing situations analyzed in earlier chapters and involves individual organizations. The other type is more complex and involves clusters of organizations. Four main types of clusters are considered: buyer clusters, seller clusters, competing clusters, and government clusters. The purchase of projects under conditions such as described here is a complicated matter. Several parties such as the end user, consultants, and contractors are involved on the buyer side. Such parties represent a buyer cluster. Complex projects consist of a large number of different products and services, supplied by several units that may either be separate companies or belong to the same TNC. These sellers form different seller clusters that compete with one another (competing clusters). Various governmental organs are often involved as a government cluster, as well, either as part of the buyer cluster or as a group of outside parties. Individual clusters are connected or held together, both internally and with each other, through four main types of linkages: product linkages, information linkages, social linkages, and financial linkages. The focus here will be on a particular seller cluster and its relationships with other clusters. The marketing situation involved is analyzed at the organizational level, with individuals seen as representatives of their organizations. These individuals, and the organizations they represent, are connected through contact nets. One complex cluster situation, in which seller and buyer clusters are spread out between different countries, is analyzed below. Such a market situation, which is called multinational marketing, requires the coordination of marketing efforts between various subsidiaries within a given TNC cluster.

Multinational Marketing

The marketing of international projects is an important activity for six of the European TNCs. This is a very special kind of business endeavor, requiring the extensive coordination of sales activities within the group of companies involved.

A major business activity for Food Equipment is to supply large and complex components which are bought through international tenders to large industrial projects. The company's marketing situation is illustrated in Figure 7.1. Since there are usually no main contractors in developing countries that can handle such large projects, it is usually done by any of the multinational giants in the field,

FIGURE 7.1. The marketing situation in multinational marketing.

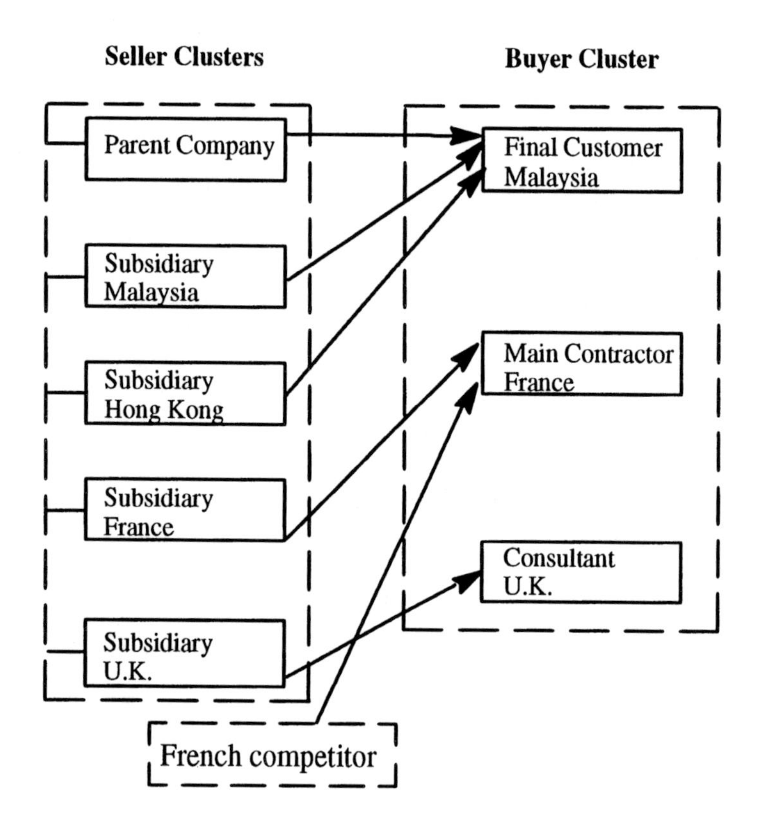

such as Bechtel. This means that most purchasing decisions are made outside the project country. One example of such a complex business operation involving Food Equipment was the building, by a major French contractor, of a large power station in an ASEAN country in the mid-1980s. The project was financed in France and run by an English consultant firm. The final customer was a large public organization in the ASEAN country. It would have been natural enough for the French contractor, which was the main contractor and also the major party in the buyer cluster, to buy French equipment. Food Equipment's goal was to sell a large and technically advanced component to the project. The company's overall actions involved its subsidiary in France, which attempted to influence the main contractor. The company's English subsidiary worked with the consultant and the local subsidiary in the ASEAN country worked with the final customer. There was also a reference project in Hong Kong which representatives of the final customer visited. Finally, experts from the company's home office traveled to the final customer to solve special problems and answer questions. Selling to large international projects of this type can thus require the coordinated actions of major parts of a TNC with its strong multinational organization. In this case there was a large subsidiary in France, which also manufactured parts to the component being marketed. The company's English and local subsidiaries were also strong and well established. The local subsidiary tried to influence the final customer to get it to adjust the tender specifications to the company's product. This was difficult. Although the final customer could determine this matter, it had to contend with the demands of the main contractor, which had total project responsibility for guaranteeing the working of the power station. That party also had know-how concerning the product in question and possible suppliers. The subsidiary managed very well to influence the final customer, which refused to buy from the company's French rival despite a strong recommendation that it do so by the main contractor.

A similar situation is faced by Tooltec which sells a specially developed processed material to international projects such as those involved in constructing large petrochemical complexes. Usually it is only the assembly which is performed locally. The sales themselves are usually agreed upon somewhere outside the user country. The buyer cluster normally consists of four main parties: the end user, the main contractor, the licenser for the process to be employed, and a consultant. It is important for the company that it have an international project group to manage and coordinate the actions of the various subsidiaries involved in marketing the product. Technical matters are discussed with the end user and with the consultant in trying to influence the forthcoming tender specification. Later, when the tender comes out and the competitors have come in with their bids, marketing changes. Even if the tender specification is predisposed to buy the company's product, competitors can copy the product and eventually take the order.

Another TNC, Weldprod, involved in selling equipment to large off-shore projects and construction projects, has had to overcome huge entry barriers similar to those already discussed. Projects tend to be dominated by main contractors from the U.S. or Japan, which favor suppliers from their home countries. It is very hard to penetrate these national clusters. Such projects can be a kind of protected industry, where a business opportunity for a particular single company outside the cluster arises only when other companies cannot fulfill the quality requirements or deliver in time. Moreover, the industrial policies of some of the ASEAN countries tend to favor Japanese firms. It is often necessary to have a local partner in the user country to be able to make bids and affect the end user. To compete for sales to these large projects, a consortium often must be formed along the same lines as those of competitors. However, such efforts are handicapped by insufficient relations with the few main contractors and the lack of Japanese-style trading houses in the home country to act as project coordinators.

The competition also varies with the economic situation. With the deep slump in the region and recession in most of the world in the mid-1980s, cutthroat competition and abnormally low prices prevailed. The number of projects went down and the number of competitors rose. An extreme situation developed in connection with one of the few large construction projects, for which an abnormally large number of companies competed. About 60 companies presented bids, and 13 prequalified.

THE LINKING PROCESS

Information costs, bargaining costs, and enforcement costs are the main transaction costs affecting how a solution to a customer need is marketed. The first stage of the marketing process to bridge the gap between demand and supply is called the establishment stage. This stage, which hopefully ends with the implementation of an agreement, was shown to follow a certain course that could be divided into five main substages, each dominated by one of the main types of transaction costs mentioned above. The different substages of the establishment process, grouped according to the type of transaction cost which dominates, are shown in Figure 7.2,

FIGURE 7.2. Stages of the linking process.

The Establishment Stage	Scanning Substage Approach Substage Bidding Substage	***Information Costs***
	Negotiation Substage	***Bargaining Costs***
The Habitual Stage	Completion Substage Follow-up Substage	***Enforcement Costs***

together with the habitual stage, which follows. In the habitual stage, the relationship has already been established and routinized, and habits regarding it have been developed. The relationship functions then in accordance with certain customs. One important task at this latter stage is to maintain the established relationship. During the establishment stage, five separate substages or phases can be distinguished: the scanning phase, the approach phase, the bid phase, the negotiation phase, and the completion phase. This chapter deals mainly with the establishment stage in the selling of projects. In addition, certain attention is directed at the follow-up phase, the final part of the second or habitual stage.[1] The marketing activities during these various phases or substages vary regarding such project characteristics as project size, purchase form, and familiarity with the seller. This book focuses on the scanning, approach, and follow-up phases. The other three phases, which cover bids, negotiations, and completion (project management), have been extensively covered in the literature.

The first three substages of the establishment stage concern mainly information costs. During the scanning phase the seller surveys different organizations that initiate projects. This is primarily an information collection task. During the next phase, that of approach, the more outwardly-directed and canvassing activities of communicative nature begin. The seller now knows of different potential projects and project purchasers and tries to influence them. At this stage the different parties do not yet know much about each other and thus experience high uncertainty. As for the buyer, this uncertainty may concern the seller's production capacities and solvency, as well as its propensity to keep agreements. During this phase the seller may help the buyer perform various studies. In this phase, the marketing activities regarding the object of sale begin in a general way and later become more specific. During the bidding phase, in turn, the seller prepares the bid on the basis of the tender. During these three phases the parties to a deal collect information about each other concerning different aspects of the need in question and how to solve it, such as actual price levels and how to compete with price in relation to quality. Considerable resources are spent in obtaining first-hand knowledge regarding the prices. These information-dominated phases end with the bid phase, which closes with a formalized offer made by the seller.

Bargaining costs dominate the negotiation phase but are also found at the previous three substages. During the negotiation phase, the sellers are familiar with the need in question and the buyers are familiar with the various offers. Bargaining takes place in order to reach an agreement. Bids are evaluated in both technical and economic terms. Bidding usually starts with a preliminary bid followed by informal meetings to specify the final offer. After that, negotiations become more formal, until the buyer finally decides on the suppliers (Ghauri 1983, chapter 3). The agreement is influenced by the state of the broader institutional framework. If for one reason or another the legal system cannot be made use of here, stipulations not normally present in business contracts may be included, or a less formal agreement may be made, based on trust, with the aim of reducing transaction costs.

The deal is implemented in the completion substage, an administrative phase. Most costs at this substage are related to enforcement. This phase starts with the selection of a particular supplier. A temporary organization is created for completion of overall assignment. The different parts of the project are delivered, installed, and put into operation. How construction, training, project management, and service are organized affects the size of enforcement costs.[2]

Finally the company reaches the follow-up phase of the habitual stage. This phase is highly important from a marketing standpoint.

It is here that companies can take advantage of spin-offs from the project. Software, for example, may be sold as service contracts. The buyer may also supplement the previous order or buy other related equipment. In addition, the seller may utilize the recently completed project as a reference for future business. In any case, it is important to maintain contacts with the buyer to obtain information about new projects.

A primary aim is to reduce the overall transaction costs for the whole marketing process. There tends to be a trade-off between the costs of information, bargaining, and enforcement. High information costs initially, for example, may be compensated for by lower bargaining costs thereafter. Similarly, when the relationship becomes more routinized through the execution of the agreement, this may reduce enforcement costs. The economics of transacting thus vary with the marketing process. Increasing the parties' knowledge of each other and of the transacting process makes it increasingly possible to organize transactions efficiently.

The most crucial marketing problems in selling to projects are often found in the early phases of the establishment stage, even before it is possible to identify what projects will be available. The seller tries to obtain first-mover advantages as early as possible. The seller's own commitment increases, and with it the size of the linkage-specific investment, as more and more substages are passed. The longer it takes to achieve a corresponding commitment on the part of the buyer, the riskier the situation becomes for the seller. A very important decision then is whether to withdraw from the relation if the buyer cannot commit or if some competitor obtains a definitive first-mover advantage. It is thus important to make a strong marketing effort early to get a lead over competitors. In that way more is also learned about the business situation and about how to act in the various phases.

The linking process will be illustrated through indicating how first-mover advantages can be attained through producing an increase in information impactedness during the marketing process. It will also be shown how first-mover advantages can arise through the establishment of social impactedness through the development of contact nets.

The Scanning Phase

As was mentioned earlier, it is important to obtain information about a project as early as possible to start influencing the buyer cluster. As in the case of Food Equipment, an information system for surveillance of potential projects can be organized at a TNC's headquarters. Food Equipment formed an information net encompassing a variety of major organizations, thereby obtaining first-hand information about emerging projects. This net involved the World Bank and its regional sister banks with all their global, regional, and local units. Similarly, it involved the U.N. and various of its organizations, different home-country organizations engaged in assistance to developing countries, and various embassies and foreign missions. Effective surveillance demands regular visits to the diverse organizations concerned, as well as an organization for receiving persons from foreign missions and other important visitors, perhaps at the central head office or at subsidiaries. Such activities can either be carried on internally by the parent company and different subsidiaries, or it can be managed externally by consultants and trade offices.

One way for a company to take the initiative at as early a stage as possible is to create a project for a customer and to then try to influence the latter to accept it. That company would require considerable knowledge of the country to know which projects could be in demand.

The Approach Phase

The two parties which it is above all important to influence in the approach phase are the end user and the consultant. For Telecom the basic marketing procedure in the approach phase is initially to try to affect the tender specifications. To facilitate this, the company endeavors to be on the spot long before a tender comes out, even before it is known when the tender will be made. The aim is to recognize what products the firm could supply in terms of the specifications. To achieve this, visits are made to the end user's technicians, and a contact net is created. Certain persons are employed to make trips for scanning purposes, and these contacts are maintained for many years. This is particularly easy for an old and experienced company such as Telecom, which has sold its systems to a limited number of customers throughout the world for many years. Usually, there is only one customer per country and no consultants are involved, and there is virtually always someone from the TNC visiting a customer. The company prides itself in having broad and continuous contacts of a positive character with previous, current, and potential buyers.

It is important to learn who writes tender specifications. These persons or groups must be visited and supplied with information. A specification for a system is very complex, requiring much outside information. This makes those writing them receptive to visits from both the firm and its rivals. In one case, the company knew very well which persons from rival firms visited a particular customer. It had been operating in the country for many years and had a long and excellent relationship with the customer. The continual and broad contacts were useful. The system that was finally sold was well-known, since earlier attempts had been made to sell it in that country. A crucial advantage was won when the company managed to exchange some old equipment for new with the customer. By means of this reference equipment, the company got "a foot in the door."

When selling equipment to power projects, Indpow tries to influence tender specifications by working with consultants for industrial projects and with technicians of the power companies involved. It takes a long time before attempts to sell projects result in an order. The company has worked with a potential customer in one ASEAN country for three years to build up its technical know-how and arouse interest in the company's products. This is very much a matter of selling ideas. When the tender is out, all the competitors submit bids as well. Thus, the aim is to make early contact in an effort to get the specification adjusted to the company's products. Since competitors act in the same way, the task is not an easy one. Some customers are more easily influenced than others. For instance, one major customer in another ASEAN country is very strict in determining tender specifications and does not allow itself to bias these toward the company. In such cases, it is best to employ more traditional sales arguments such as emphasizing the company's long commitment to the country, its established technical center, and the high quality and early delivery of its equipment.

Paving systems introduces itself in a new Southeast Asian market by arranging special seminars for interested key persons within the buyer net, chiefly public authorities, foreign as well as local contractors, and consultants. It is important to have good relations with consultants when establishing business in the area. The company's first order was won very much due to a consultant's recommendation. The next order it received was for a project financed by the World Bank. The person in charge of the project at the Bank worked out the tender specification together with a consultant. The specification was written largely with a bias toward the company's system, a result of good contacts with both these parties. The bias was nevertheless softened sufficiently to make it possible for a competitor to make a bid. The company tries to get specifications to favor its products even if they are written in a seemingly neutral way.

When one of Food Equipment's subsidiaries introduces a product to a large project involving a major public customer, it is concerned with many of the customer's employees from various departments. Some of them are key persons, more important than the others. Food Equipment knows who they are and tries to influence them. This occurs not in meetings of the tender evaluation committee, to which various of these persons belong, but elsewhere, in their offices, for example. It takes a long time to build up enough confidence to get the type of information desired. It is a delicate act of balance involving judgments of how much one can ask and how far it is possible to go in trying to influence the person.

Consultants are often an essential party in the writing of tender specifications and it can be important to establish good relations with them, too. They frequently make feasibility studies on which specifications are based. For World Bank projects such studies are mandatory. To get a first-mover advantage at a very early stage, the seller may decide to prepare a feasibility study itself, thus bypassing the consultant. Such studies, on the other hand, are very expensive to make and there is, of course, no guarantee of getting the order. Conmine, for example, made a feasibility study for a large construction project, a study which was then supplied to all the bidders.

The Bid Phase

The bid phase starts with the buyer publishing the tender and the seller submitting a bid. A critical decision in this context is how much to invest in a bid, that is to which extent the seller should become further involved in the relationship. The stakes are rather high since it is expensive to produce bids, particularly for large projects. The chances of getting an order depend on how much the seller has managed to tie the buyer in previous substages.

The Negotiation and Completion Phases

Sometimes the first part of the negotiation phase, the evaluation of the bids, is open rather than closed, making it possible to influence the buyer. Under these conditions the bids are usually short-listed and a few sellers are invited to discuss various types of clarifications. Such meetings tend to be extremely important. Since not everything can be formulated on paper, oral specifications may be decisive for the outcome of negotiations. Technical matters often dominate.

Final negotiations of the contract for large and complex projects often take place even after an order has been accepted. For one of Telecom's projects negotiations continued regarding the commercial stipulations, although the technical stipulations had been agreed upon earlier. Financial issues were also negotiated at this late stage for the first time. However, such agreement conditions are normally negotiated earlier since they are usually stated in the tender specification. A final decision on prices tends to also be made in this final stage. Then the project is implemented.

The Follow-up Phase

The sales involved in a project are not usually over when the contract has been signed. The spin-offs generated are often considerable. Also, even if a company has lost a contract, there are sometimes possibilities for sales in the follow-up phase.

For Paving systems the machines delivered to a factory generate follow-up sales of chemicals to the customer equal in value annually to about a third of the value of the machines. Thus, the company tries to bind the customer to the purchase of both through a combination offer. The guarantee for the machines may be formulated in such a way that it is no longer valid if chemicals from other suppliers are employed.

When it won the contract for its system in one ASEAN country, Telecom shut out all competitors except one from further orders. To be secure, the customer wanted to have two suppliers, but it considered having three or more suppliers too complicated. Since a rather large volume was still to be ordered, Telecom's first-mover advantage was very important. The buyer informed it and the other supplier that the company which best managed the completion of its particular project would get the supplementary orders.

Besides being able to sell additional major equipment and service, Telecom also received several orders for supplementary equipment. In addition, a joint venture was later formed with a local partner for building and managing a part of the overall system. This also meant the delivery of more equipment.

All these follow-up sales demonstrate how an initial order for main equipment can be worth much more later. This is often taken into account in marketing strategy through the selling of basic equipment at a low price in order to gain access to a profitable follow-up market.

THE LINKAGE STRATEGY

Linkage strategies concentrate on social linkages being created and maintained with various key persons. For those from the TNC (the focused seller cluster), it is necessary that they know about new projects and about other seller clusters, in particular about how these sellers are evaluated, how they influence decisions, how they follow up on contracts, how they are paid, and how they deliver and install projects. The establishing and maintaining of contacts often represents the largest and most important part of the marketing investment for project selling.

For Paving systems, personal contacts are the most important factor at every project phase. Letters and telexes work only in exceptional cases. Without personal contacts, nothing can be achieved. Especially important are the contact nets relating to the tender specification, which several persons take part in establishing. In the case of the first system sold, all decisions on the customer's part were taken by the Chinese director of that company together with his three sons. The contact net to obtain the next order was more complicated. It included the managing director of a public roadbuilding company and three of his subordinates, the regional representative of the World Bank, and the regional director of the World Bank in Washington, together with a consultant which the World Bank had appointed. The firm learned of this project through a marketing survey of its own, in which visits were made to the consultant and to the regional director at the World Bank. It took two years from the time of the first contacts until the order was received. A third order came two-and-a-half years after the initial contacts. Lead times in the selling of projects thus tend to be long.

For Indpow personal contacts with ASEAN customers, especially in the case of local Chinese customers in the private sector and for the most part in the public sector as well, are very important, much more so than with customers in northern European countries. ASEAN customers only do business with those they know personally and trust. Thus it takes a very long time to establish relations with them. The contact pattern with transnational firms in the area is somewhat different. Expatriates are more involved there, giving dealings more of a European or North American flavor. Initial contacts with ASEAN customers generally involve both management and technicians from both parties. It is highly important in marketing one's products to create and utilize various contact nets. Contacts are specific, take a long time to build up, and should be personal. A large portion of business contacts take place after office hours.

Indpow's personal contacts with a large public organization in one ASEAN country are also particularly important. They are maintained at several organizational levels. Approximately 25 persons from the company have contacts with some ten persons at higher levels and 35 persons at lower levels of the customer business. Differing types of information can be obtained from these contact persons, depending on how close contacts are with them. Although competitors work in the same way, Indpow is better established. Contacts are the biggest and most important part of the company's marketing. It is therefore essential that contacts be handled in the correct way. If a conflict with a contact person should arise, the relationship would be very difficult to repair. One might well have to wait until the respective person on the seller or customer side had left or been replaced.

For large projects, it is important that persons high in the organization establish contacts with persons high in the customer organization to create status and show that the deal in question is important to the seller. It must be apparent that the Managing Director of the subsidiary participates in the decision process. At certain stages in the final rounds, high-level representatives from the company's central offices or from a large product company or division in Europe may join in. Seniority, as already indicated, is essential in the region.

THE COMPETITIVE STRATEGY

In the selling of large projects, Indpow considers its local organization and its reference projects to be highly important. Its financial strength and technical ability are crucial as well. The latter two factors limit the number of competitors which can enter, since only companies strong in these areas can fulfill the tender specifications.

Critical issues in this context are where to locate and how to coordinate the various engineering capacities needed to offer large, technically sophisticated products and projects. In most cases, the engineering capacities required are located in several units, such as in a sales subsidiary, a regional unit, the product company, or a special research and development company. Indpow has an engineering unit in a central location in SEA. This unit produces all the offers and is responsible for all of those offers within its field for the markets in the region. Very large and complex projects are coordinated with the product company or with a special project group located in Europe, since the companies located in the region cannot manage this on their own. For sales of technologically simpler equipment, the engineering expertise needed is often located at the individual sales company level, giving the company control of the tendering process for the products in question.

In one of the ASEAN countries it is more difficult for Indpow to influence the tender specifications. The performance characteristics there are specified by the customer, and sellers decide on their bids from that. The evaluation of the bids is then based strictly on the technical performance level specified in the tender. This applies particularly to the country in question and tends not to apply to the same extent in other ASEAN countries. It favors those companies that can produce equipment up to the technical performance level although this is no guarantee of their success. A good example of this can be seen in the evaluation of the bids for one very large project. The technical performance level of Indpow's bid was above the tender level, whereas a Japanese competitor's bid was below the tender level. Such a situation was shown to favor the Japanese company because of a large difference in price. Indpow offered much more transfer of technology and also local production of the equipment in a joint venture with a local partner. These advantages were included in the evaluation and were quantified. The price differential in comparison with the Japanese company, however, varied between 6 and 12 percent, depending on how the calculations were done. Since this was above the 5 percent differential stipulated in the customer's guidelines, the Japanese company got the order. It was thus better here for a company to offer below the desired performance level but have a low price than to provide above that level and to have a high price, despite the obvious advantage of a joint venture with a local firm for the manufacture of most of the equipment and for the accompanying transfer of technology. Such a buying strategy on the customer's part favors those sellers that can adapt the technical performance level of equipment, and thereby its price, to shifting customer demands. This is not an easy task and is often not possible for the TNCs studied, the products of which are characterized by rather uniformly high technical levels and resultingly high prices. Japanese corporations are much better at this strategy. They can offer products which are better adapted in this respect at lower prices, and are thereby more sensitive to local customer demands. Hence, the most important marketing variable after fulfilling the tender specifications is price. This is something which is generally valid for the entire geographical area.

The importance to the company of financing varies with the size of the project and the country of origin. In one of the countries it is an insignificant factor in any case, in another it is a principal marketing variable in the case of large projects, and in still a third country it is always very much of a factor. Unfortunately, the TNC cannot offer the same highly favorable conditions in this respect as the main Japanese competitors.

As the case cited above suggests, it can be difficult to know in advance how a buyer will appraise such a factor as the transfer of technology. Calculations of the value of this competitive advantage are also arbitrary, making use of this marketing variable rather unpredictable. For large projects of one particular type, however, it is a critical component in competitive strategy. This has induced Indpow to establish a technical center in the country, in which the demand for such projects is largest. This is a way of coming closer to the customer and of circumventing the effects of the long distances between Europe and these markets.

The most essential elements in the competitive strategy for systems which Paving systems sells in the private sector are quality, transfer of technology and financing, and availability of local service. Unfortunately, the company cannot offer any competitive financing here, a weakness it shares with most of the companies studied. Financing is not as important a factor for public customers, since their projects are often financed by the state budget in the respective country or by loans from the World Bank. Price and times of delivery are less important as well in the public sector. The necessity of a good reference project for success in selling to projects, however, cannot be stressed enough. The first project sold worked very well and was employed as a reference object. It has been visited by several potential customers and was a major reason that Paving systems won two later tenders.

The competitive situation changes after suppliers have been selected. Only those companies that received orders are left and they have clear first-mover advantages over the others. In one case in which Telecom got an order, the most important marketing variables for orders in the follow-up stage for the two remaining TNCs were delivery times, installation, and after-sales service. One of the companies was considered to have the upper hand for the latter variable and the other to have the upper hand for the first two.

CONCLUSIONS

A major finding of the study in the project marketing area is the existence of a need for long-term relationships. For narrow customer specialists the degree of linkage-specificity is high. In the case of trilateral governance here, formal agreements are supplemented by the strengthening of relationships between the parties. The transaction cost model presented in Chapter Four is also applicable for analyzing such marketing problems.

The marketing of projects is very much a question of the selling of ideas, since during a large part of the marketing process projects are only found on paper. The product linkage thus comes in very late in the process. The essential influence of the customer cluster is based on the information linkages and social linkages. A crucial part of marketing investment is in the establishing and maintenance of contact nets. Permanent contacts with buyers are necessary in order to influence decisions, to be paid, to carry out follow-ups of various decisions and to get the project delivered, installed, and put into operation. The contact net concerning the tender specification is of utmost importance for success in this type of marketing.

Marketing strategy aims at always having the right competitive and linkage strategy in order to obtain a first-mover advantage over competitors. The main variables in a competitive strategy tend to be quality and the transfer of technology, whereas customers tend, in fact, to chiefly favor price. This can result in difficulties, particularly in competition with Japanese TNCs.

The high linkage-specificity and long-term character of relations in selling to large projects make the losses to any of the parties large if anything should go wrong with a project. The seller's investment during the initial marketing phases is often substantial, and this can usually not be recouped if an order is not received. This uncertainty can be reduced through safeguarding the investment with the help of a third party or through establishing a reciprocity in the parties' gradually increasing involvement in the relationship. Particular marketing problems are often encountered in the early phases of the marketing process, even before it is possible to identify individual projects.

The linking process can be divided into six phases: the scanning phase, the approach phase, the bid phase, the negotiation phase, the completion phase, and the follow-up phase. During the scanning phase the company can create an information collection network, at least in connection with the most important projects, thus generating contact organizations and individuals. During the approach phase the company's marketing efforts are mostly concentrated on influencing tender specifications in the direction of the performance characteristics of its own products. If this is impossible, other marketing activities may be undertaken. Since the follow-up phase can result in large complementary orders, marketing should not stop with completion of the project. Rather, the follow-up phase should also be viewed as an element in the company's overall planning of the marketing strategy for its projects.

Multinational marketing in connection with international projects is a difficult task, since it requires considerable coordination of the activities of different units within the TNC and of various parties within a rather complicated buyer cluster. Since the companies from a given country tend to stick together, there is often competition between different national clusters. A national cluster is very hard for companies from other countries to penetrate when both the buyer cluster and the seller cluster come from the same country.

NOTES

1. See, e.g., Bergström (1980), Lindberg (1982), and Ghauri (1983) for discussion of the establishment of such business deals. These authors divide the process, however, into other stages and give them other names.

2. See, e.g., Hadjikhani (1984) regarding problems of manpower training encountered in international projects.

Chapter Eight

Implications for Industrial Marketing Management

Whereas the previous chapters were mainly of descriptive character, this final chapter is prescriptive in its emphasis, providing concrete advice for industrial marketing management on how products should be marketed in Southeast Asia. In this chapter, we are concerned with strategic management and marketing management in that particular area. The concrete advice should not be interpreted as applying to management or to international marketing generally, in view of the vast differences between markets and countries. Also, reality even in that geographical area is far too complex to be captured in simple models and simple prescriptions of appropriate behavior. This book merely presents a challenge to the practitioner, who after having learned the language of the approach to industrial marketing presented here and having entered into the world of transaction-cost economics, can see industrial marketing problems from a new perspective and make use of certain new analytical industrial marketing techniques. Also, the reader is provided access to a broad fund of experience on the marketing of industrial goods in Southeast Asia through the numerous examples of how different transnational corporations market a wide variety of products there.

A major message of the book is that although there are many commonalities between the industrial marketing approaches of these firms, particularly at the strategic level, differences also abound. Details of how to market an industrial product vary. A practitioner may thus find some parts of the book much more relevant than others depending upon how close they compare to actual experience. Ideally, industrial marketing managers or other interested readers will take note of the experience of the companies reported here, in particular in Chapters Three through Seven, and consider what it has to say in view of their own situation. For example, as observed in Chapter Five, the need for sophisticated market segmentation differs from one company to another, based among other things on the company's strategic marketing profile. One particular company employed rather advanced culturally based segmentation which resulted in a detailed clustering of customer organizations. If the reader is associated with a company, for example, which has a definite need of such segmentation, cases presented in that chapter may serve as a prototype for company plans to establish itself in such markets and improve its current segmentation strategy. For a reader associated with a company with no such need of segmentation, such cases in that chapter may be of no direct prescriptive interest and simply provide insight into how companies existing under quite different circumstances behave. Alternatively, this same reader may see the case material as pointing to the general importance of cultural segmentation.

It is therefore evident that one should be cautious when drawing conclusions for industrial marketing management on the basis of the findings considered here. The importance of a broad approach to international marketing should be stressed again and again. Management should beware of being myopic. It is vital that one not become so engrossed in how marketing has been carried on earlier that one loses sight of what the presence of a differing environment may imply. An institutional approach can be very helpful in analyzing the specific, contextually dependent conditions that can affect marketing. Both economic and noneconomic institutions are important in their effect on marketing based on transaction costs. Such institutions constitute a framework for industrial marketing.

Governance forms are broad categories of how transactions are organized. These infrastructures constrain marketing strategies, in a sense marking the playing field. In less industrialized countries, in particular, it is very important that a company that wants to establish operations start, before marketing is begun, by determining the presence, status, and basic functioning of different economic institutions in the country and the geographical region involved.

Transaction costs represent the driving force behind the economic organization of marketing in Southeast Asia. They point to factors of interest to strategic industrial marketing. The efficiency motive is of overriding importance. Greater efficiency can be achieved by reducing any or all of the three basic types of transaction costs, those of information, bargaining, and enforcement. However, it is up to the practitioner to give these broad constructs specific meaning for a particular marketing problem or issue of interest.

The critical importance of relationships in industrial marketing in Southeast Asia cannot have escaped the reader. Thus, the marketing of industrial goods is very much a question of building and maintaining linkages. The degree of linkage specificity is important since it indicates the seller's dependence on specific customers and suppliers, and their dependence on the seller. This concept is dimensionalized in four different ways in the model presented here: as product specificity, as investment in asset specificities, as degree of impactedness, and as degree of substitutability.

Keeping these dimensions in mind, the manager of a company can design a practical marketing strategy consisting of a linkage strategy and a competitive strategy aimed at providing the company first-mover advantages, for example. This can be done for each of the various stages of the linking process and in accordance with the conditions and strategic marketing profile specific to the company.

LINKAGE STRATEGY

The matters discussed in the individual chapters have many implications for industrial marketing management. Particularly in Chapter Five, the significance of contact nets becomes clear. Contact nets are one form of linkage around which everything revolves in the marketing of advanced industrial goods. In Chapter Five I indicate how contact nets are established and sustained in a linkage strategy.

The distinction between two basic principles of market segmentation, namely segmentation between organizations and within organizations, is important. The first principle, segmentation between organizations, represents holistic segmentation. It was illustrated for several companies, Pacmat's cultural grouping being the most elaborate illustration. The experience of this company in connection with managing the contact nets to different company groups also illustrates the importance of knowing the host country and its culture well. This makes it easier to differentiate various segments on the basis of the behavior of customers and to discuss various matters raised during meetings. Segmentation within organizations is characterized as representing embedded segmentation. This principle was shown to be vital for companies that are dependent on a few large customers. The two basic principles, to be sure, are possible to combine directly, although they can best be seen as complementing each other. Price behavior is a vital variable in SEA which leads to segmentation both between and within organizations.

Successful industrial marketing in SEA is intimately associated with the management of social linkages. A vital element here is trust (see Figure 3.1). Relations are more person-oriented than they are in Europe, for example. To a high degree trust in SEA is based on personal traits, where membership in the same social/cultural group, often that of the Chinese, clearly facilitates trust. Professional trust, which has to do with how well a person represents his company and what the company stands for, is not as dominant as in Europe. This makes individual trust relatively more important than organizational trust. Organizational trust is also frequently a consequence of individual trust. Brand loyalty, accordingly, is lower. One consequence of the greater emphasis on individual trust is that there is greater risk of opportunistic behavior. An example is given in Chapter Five of a supplier who was squeezed for better terms despite having had a long-term relationship with the customer. Another consequence of the emphasis on trust the frequent haggling, which can also be explained by the way that purchasing is organized, with bargaining as a part of the purchasers' formal evaluation of the seller.

Networking is hard to plan in SEA. Individually based relations must be allowed to evolve to a certain degree on their own. Persons should meet to discover if any liking for each other develops. Certain contacts are more or less secret and are highly personal. The results of the study imply, however, that this is an area where considerable management improvements can be made. Meetings can be more consciously arranged. Moreover, better use can be made of information systems, for instance a computer can help to map contact nets. Also, when an employee leaves the company, it is important to retain as much of that person's contact network as possible.

Social linkages are hard to manage. If something goes wrong

with sales management, sales representatives may leave the company and take customers with them. There is also a considerable cultural barrier. The salespeople are mostly Chinese, and they behave differently from Westerners, something for which sales management must account. The business language is also largely Chinese, and *guanxi* is strictly for the Chinese. An expatriate sales manager from the West may be unable to control salespeople in the East. Control is rendered even more difficult by the kickback systems in some countries. The recommendation here, both from an instrumental and a moral perspective, is to stay out of it and take more of a long-term view. The risks in taking part in kickbacks are too great. Such a view may well be considered naive by some more practically oriented businesspeople, who feel that it is impossible to do business in certain industries and countries without becoming involved in kickbacks, which they view as a kind of surcharge. However, this issue cannot be taken lightly by top management, particularly at the headquarters, due to the risk of international repercussions. Business is global today and so is a company's reputation. The question of how to deal with the risk and temptation of kickbacks is thus of vital strategic importance and should be treated as such.

The various circumstances just discussed make the process of becoming established in SEA a long, drawn-out affair. As shown again and again, relationships normally take considerable time to develop. Personal selling here is more than a promotional tool, since specific individuals must create equally specific linkages. The implementation of a competitive strategy rests mainly upon those individuals.

COMPETITIVE STRATEGY

The key issue of competitive strategy in Southeast Asia, as analyzed in particular in Chapter Six, is how to make an efficient trade-off between quality and price for the specific types of markets there.[1] As evident in that chapter, product adaptations were rare among the firms studied, which were marketing products largely produced in Europe. These firms tended to have simply transferred to SEA the same basic marketing strategies that they used in other parts of the world, strategies emphasizing the high technology and quality of their products. This results in a competitive disadvantage in these very price-conscious markets, where the quality level of their products appears to be much higher than that which customers expect and demand. The concept of reliability as it is understood there can be used to illustrate this. A "reliable" product in such markets is basically a product of fairly low cost and of a quality of about the same level as that which the market demands. The quality of a product is taken here to include both the product itself and its maintenance and service. Results of the study suggest that it is the Japanese companies, in particular, which excel at such a strategy. Market research is important for obtaining insight into customers in such respects and for determining the appropriate quality level. Japanese companies are known for their excellent marketing intelligence networks generally and also within SEA.

A critical strategic issue addressed repeatedly in this book and shown in Figure 8.1, concerns price versus quality. To examine this matter further, consider first a market situation approximating that of market governance. Assume that the companies involved operate in one single market and sell comparable products which cater to rather simple needs. Assume as well that production costs are constant. The curve shown in the figure indicates a general association between the competitive parameters of price and quality. A particular combination of the values of these two parameters leads to a given company's earning a certain profit, in that the combination of price and quality in question is coupled with a specific level of demand. Let us say that point P on the curve represents the most profitable combination. At this point P, transaction costs are at a low level. Companies discover a market-efficient price by comparing different prices in terms of their effect on profit. This is fairly easy to do, since most information of relevance to the product is contained in the price. However, there is also a need to collect information on market demand for the product and to bargain about the price until a price level is reached. The parties also make agreements, which reduce enforcement costs as much as possible.

The main transactions observed in the present study, on the other hand, take place through relationships, which require extra market exchange involving bilateral or trilateral governance. Due to these

FIGURE 8.1. Price versus quality.

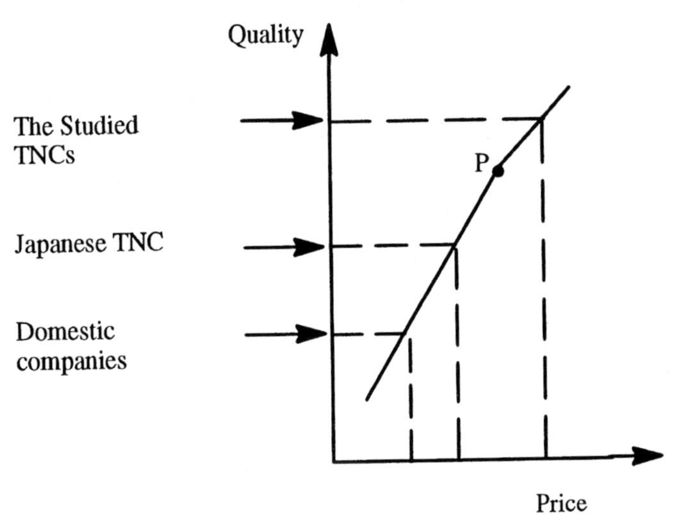

governance forms, the situation does not conform with that in Figure 8.1. Accordingly, the price at point P, under the present conditions, reflects the value of an efficient resource transfer within a linkage, a transfer which results in high linkage efficiency for both parties. Only those parties involved in relations that reach a certain efficiency level survive. In this situation of theoretical equilibrium, sellers can demand the highest price compared with the transaction costs incurred. Any deviation from this point results in a lower profit, either through higher costs or lower prices. Japanese firms can be assumed to be situated closest to P. They have found the most efficient combination of quality and price in relation to a specific need for a certain combination of transaction costs. The European TNCs studied can be assumed to be located at a point above this on the curve and local firms at a point below it. The quality level of the products of the firms studied is closer to the ideal point P than is the price of the products. The opposite characterizes the local companies. On the basis of this abstract analysis of the situation, it would appear that the Tics studied should decrease both quality and, in particular, price. Adapting their products more to local markets could pave the way for the necessary price reduction. At the same time, the analysis above suggests that it would not pay for them to reduce quality and prices too much in an effort to approach the market conditions applying to local firms, since this would result in a competitive disadvantage. The local firms, in turn, would apparently do well to increase the quality of their products, even if this means a certain increase in price as well.

At the same time, the matters discussed in Chapter Five suggest that Figure 8.1 simplifies the market situation in Southeast Asia too much to provide an adequate overall view of the functioning of the market there. For companies operating in terms of bilateral and trilateral governance forms, it is much more difficult to determine linkage efficiency. Uncertainty is high when such forms apply, making it virtually impossible to determine in a specific way the needs there. Because sellers markedly influencing customer needs, it is difficult or impossible to determine when the point at which the transaction costs incurred give the best price has been reached. This reduces the efficiency of resource allocation and it makes it hard for price makers to hit the mark in their efforts to attain the best price. Such a mark also becomes a moving target.

To make the situation depicted in Figure 8.1 more realistic, one can assume that this figure only represents the one segment of the market, albeit the largest one, in which Japanese companies dominate. The figure indicates at what a disadvantage other groups of companies are in this segment. Their strengths lie in other parts of the market. The Tics studied are more efficient in the high-quality segment, whereas the local firms are more efficient in the low-quality segment. If other figures were drawn to represent these submarkets, Japanese companies could be shown to be at a disadvantage there. The group of firms which dominates a submarket has managed, therefore, to achieve a better approximation of the ideal price and quality than its competitors, giving them lower transaction costs. At the same time the efficient combination of price and quality they achieve shelters them from other firms through the high degree of linkage specificity between them which can thus develop, giving them a particular competitive advantage in this respect. As the figure indicates, however, this gives them no permanent refuge from competitors. Competitors can in time expand into this and other market segments, their ability and readiness to do so depending to a considerable extent upon the switching costs this would involve.

An important question for the companies studied is thus whether they should expand more into the Japanese medium-quality segment. The low-quality segment is mostly out of their reach because of the high transaction costs entailed in the learning, bargaining, and deal enforcement required to adapt their high-quality products to such a market segment. This would also mean leaping over the market barrier involved in going from bilateral/trilateral governance to market governance. These companies would also need to consider what such a strategy would imply for their high-quality image in their main European and North American markets. Their products are simply not suitable for the tough price competition prevailing in this low-price segment. Their organizational costs here would also be too high to allow them to earn a profit. One way to express this is in terms of high transfer costs. Even expansion to the medium-quality segment, although it would largely involve the same governance form, would be difficult. As is evident in the book, products developed for advanced industrialized countries are expensive to adapt to less industrialized countries. The information costs involved in learning about that submarket are high, as are the bargaining and enforcement costs such a market would require. To these transaction costs would be added the production costs needed to adapt the product to this market. Through simply copying the marketing strategy employed in its main markets, the company constrains itself to a narrow high-quality segment, a particular marketing strategy, and a particular organization of its marketing activities. It is indeed as difficult for competitors to break into this market segment as it is for the Tics studied to leave it and break into other segments. The key limiting factor here then is the cost of product adaptation. At the same time, there is a danger in getting stuck in one single segment. If the segment is too small, this can be a competitive trap.

Japanese firms usually have a competitive disadvantage in the high-quality segment, a disadvantage manifesting itself in particular in their unwillingness to transfer their technology to customers. The costs for enforcing the agreements this would require would be too high. They need to avoid the loss of their key technologies which would promote the development of new competitors. This drawback seems to be deeply embedded within Japanese organizations and can be traced to certain aspects of Japanese culture. The latter is highly ethnocentric, which tends to aggravate other groups and to reduce cooperation between Japanese businesspeople and people from other cultures (Stening and Everett 1984; Pye 1985, 286-91). Human resource management thus seems to be an Achilles' heel in the long-term investments of Japanese companies in SEA. On the other hand, their long-term investments in the building of contact nets with politicians and administrators are largely a competitive strength.

The seller and the buyer may differ in their views concerning needs and how these should be met. As was shown in Chapter Five, the not very technically advanced customers common in LICs tend to demand rather simple, standardized products. They also overemphasize price. This reflects the high degree of competition in the markets there and the low inflation rate, together with the price sensitiveness which is a strong trait of the local business culture. Local firms in SEA tend to be rather traditional in their views on how markets work, even if they operate in markets characterized by bilateral governance, involving complex needs and products which are more specialized than standardized. In such cases, when customers view themselves as operating more under market governance than under bilateral governance, it is difficult to introduce advanced products from other markets which cater to special and complex needs. Those markets may not be ready for certain types of products. Major questions for Tics in marketing their industrial products in LICs thus concern when and how rapidly industries which demand certain types of products will grow.

The major difference in pricing behavior between European and Japanese Tics considered above has been corroborated in another study. Lecraw (1984) reported the pricing strategies of 153 subsidiaries in the ASEAN countries between 1977 and 1979. In introducing new products, 63 percent of the Japanese firms, 23 percent of the European firms, and 23 percent of the U.S. companies priced the products below full cost. Some 22 percent of the Japanese companies, 6 percent of the European firms, and 8 percent of the U.S. firms even priced their products below direct cost. Japanese firms also introduced older and less sophisticated products into the market than did their Western competitors. Market penetration therefore was found to be a primary competitive strategy of the Japanese, whereas Western firms emphasized unique quality advantages. The use of predatory pricing (a kind of dumping) in the introduction of new products, though common generally, was found to be particularly widespread among Japanese companies. There were also indications of price discrimination between different countries during that period. Oligopolistic coordination of pricing strategies through price leadership, on the other hand, was found to be lower than had been expected on the basis of market structure. However, the period of investigation may not have been typical for two reasons, first because market structure and market size have been changing rapidly, and second because Japanese Tics have been expanding their market shares in many industries.

PRICING

The present study clearly demonstrates price to be the major competitive factor in SEA. Most companies either become involved in price competition directly or they try to avoid it by emphasizing quality and related aspects. Nevertheless, price is important in connection with each of the governance forms. A major finding of the study was that, even within the framework of the highly personal social relationships characterizing industrial marketing in Southeast Asia, pricing is important. A company located in the market there must therefore be a capable price maker. A price strategy which is well planned, for example in differentiating adequately in price between various customer segments, is a necessity for profitable operations. Pricing is very much constrained by external factors in the market. Markets tend to be more competitive under conditions of lower inflation, greater price-consciousness of buyers, and greater price continuity. Price behavior cannot be understood without taking account of customers, both as regards individual purchasers and how purchasing tends to be organized. A thorough knowledge of Chinese business culture is imperative. Otherwise, the bidding and haggling, for example, becomes rather incomprehensible.

Efficient pricing by a local subsidiary of a TNC does not stop with consideration of the market. Pricing policy toward companies within the group may be equally important. Both linkages with customers and linkages with internal suppliers are crucial. Limitations in the pricing opportunities on the market may be compensated for by the skillful use of an internal pricing system.

DISTRIBUTION

Even for those industrial companies marketing high-quality products but primarily engaged in distribution, proximity to the customers is vital. Distribution is completed with the establishment of direct linkages with final users. A presence in the local market is recommended to control prices, sales, and distribution better. This can take place either through operations of one's own, as in the case of a distribution specialist, or through maintaining good relations with distributors, as in the case of a distributor specialist. If such specialists find direct linkages with final customers to be important, for example to get to know customers better or to facilitate the transfer of technology, a sales company should be located near both distributors and customers.

MARKETING OF PROJECTS

The marketing of projects represents the industrial marketing by narrow customer specialists. A major problem in this context is that of reducing the high uncertainty which characterizes this type of marketing. Such uncertainty is a function of the high degree of specialization which limits a business to only certain customers. This entails a high degree of linkage specificity, a low degree of substitutability (because assets are expensive to redeploy), consequent high switching costs. Assets are tied to a high degree with a specific customer. Product specificity is high too, as is the degree of information and social impactedness. Costs due to the establishment of certain linkages cannot be recovered if the investment does not result in an order. Along with the high uncertainty, there are also highly complex needs which require considerable resources to discover and find solutions for. A main objective in marketing of this sort should be to reduce such uncertainty. Third-party assistance may be called in to safeguard the project during its development. One kind of assistance of this sort is a stipulation in the final contract for how conflicts are to be solved, signed by the buyer and seller in accordance with national or international law. This might be more fruitfully looked upon as a marketing of contracts (agreements) than as a marketing of products. This in turn is a matter closely related to what was discussed in Chapter Seven, where it was pointed out that the marketing of projects is very much a question of the selling of ideas, as projects only exist on paper during a large part of the marketing process. The contract then becomes the final description of the project in writing, specifying the ideas, detailing how they will be carried out and put into effect, and listing the rights and obligations of the parties. Another kind of third party assistance, used mainly by the buyer, is the use of consultants, particularly to specify the needs involved and to detail the requirements for solution. Another third-party safeguard for the seller is to spread out the risk by sharing it with other parties in a consortium or sharing it with other units of the transnational corporation to which it belongs.

Such safeguards through third-party assistance are not enough to reduce uncertainty sufficiently, however. Reduction in uncertainty can also be achieved through the linkage itself. This is especially relevant in the lengthy business transactions which take place between when marketing begins and when it results in the delivery of a project. It is extremely important to establish early a position to influence the buyer and to follow up continually through appropriate contacts throughout the various stages of the linking process. A major finding of the study is that the early stages of this process–the scanning and approach stages–are so critical for marketing success. Contact nets are of extreme importance in creating such a first-mover advantage, chiefly through the creation and maintenance of trust. Social and information impactedness should be established to create a reciprocity between the parties.

The importance of linkages in the marketing of projects and the central role of such linkages in trilateral governance is another main concern. It should be borne in mind that the linkage relationships here differ from those characteristic of bilateral governance. The differences here partly reflect the time dimension. Projects are intricate solutions to highly complex needs. Linkages there are mainly established to learn as much as possible about such needs, to prepare and implement solutions, and to follow up on this. All such activities center around the occasional transaction: one single, very large business deal. In bilateral governance, on the other hand, linkages constitute a framework for repeated business. However, there is no clear border between these two governance forms. In the follow-up phase of the linking process, trilateral governance may change to bilateral governance. A sale of equipment may be followed by repeated deliveries of process materials or spares. Time is a much more essential element in trilateral governance, expressed in terms of the linking process. The secrets of success in marketing involving this governance form have very much to do with how marketing efforts are sequenced and are evaluated over time, that is at the different substages of the linking process. The significance of the tender specification in this context should be obvious to those who have read Chapter Seven. However, one should not forget the follow-up phase, which is the gateway to continuing profitable business after the main project or product has been delivered. It is not recommended that one stop here without taking advantage of the strong first-mover advantages created in the establishment stage.

It is vital to distinguish between the two types of marketing situations found in the marketing of projects since they result in quite different market segmentation practices. The main segmentation issue as regards individual organizations concerns how to combine the individual organizations into groups or to divide them into parts. When the marketing situation involves clusters, on the other hand, it is mostly a question of combining the right organizations into a given cluster and to then treat this cluster as one organization, which means dividing it into suitable parts.

International projects are very complex, not only in view of the temporal aspect, but also because competencies are spread out internationally between different countries. To the complex time dimension must also be added a complex spatial dimension. Multinational marketing requires the highly effective coordination of marketing efforts, both in time and in space, between the different units located in various countries. Transnational corporations are in a very good position to engage in business of this sort. They possess a high capacity for the internal coordination of marketing. But then they must also, to be sure, have the internal capability to achieve solutions and to deliver them. If this is not the case, cooperation with other firms, for example within a consortium, is necessary.

ORGANIZATION OF INDUSTRIAL MARKETING

Efficient coordination within multinational marketing requires the efficient organization of various activities. The organizational problems involved tend to be large both for the TNC and for the individual subsidiary. Projects are hard to anticipate. Thus, when they appear, resources are temporarily pooled in the form of project groups. The size of a group has an important effect on its efficiency. Splitting costs and benefits between companies involved in such transactions can be problematic. Since individual subsidiaries are usually organized for more regular sales, they can run into difficulties when participating in these groups, both in reaching their usual sales targets and in motivating the people involved to engage in marketing efforts not directly related to immediate sales goals.

The intimate relationship between marketing strategy and the organization of marketing has vital implications for industrial marketing management. If quality is stressed, service and the transfer of technology become critical elements in competitive strategy, which can be organized most efficiently through the subsidiaries of the organization. If price competition is instead preferred, flexibility is likewise required in the marketing organization, which for a TNC is achieved through its agents and distributors.

NOTE

1. A similar result was obtained in a study of Swedish investments in SEA (Lundgren and Hedlund, 1983, 63), which found that Swedish companies consider the pricing of products and services in SEA to be a considerable problem. It was ranked second in importance along with "the solving of problems among domestic employees." "To overcome cultural differences" was ranked first in importance. In a 1982 ranking of the most important future problems, "pricing" ranked second after "increasing competition from foreign companies."

References

Alchian, A. A. and Demsetz. 1972. "Production, Information Costs and Economic Organization." *American Economic Review* 62, pp. 777-795.

Alchian, A. A., and S. Woodward. 1988. "The Firm is Dead; Long Live the Firm. A Review of Oliver E. Williamson's The Economic Institutions of Capitalism." *Journal of Economic Literature*, Vol. XXVI (March), pp. 65-79.

Alderson, W. 1957. *Marketing Behavior and Executive Action*. Homewood, IL: Irwin.

Ames, C. B. and J. D. Hlavacek. 1984. *Managerial Marketing for Industrial Firms*. New York: Random House.

Asia Yearbook. 1990. Hong Kong: Far Eastern Economic Review.

Bagozzi R. P. 1975. "Marketing as Exchange." *Journal of Marketing* 39 (October), pp. 32-39.

Bagozzi, R. P. 1979. "Toward a Formal Theory of Marketing Exchange." In *Conceptual and Theoretical Development in Marketing*, edited by O. C. Ferrel, S. Brown and C. Lamb. Chicago, IL: American Marketing Association, pp. 431-447.

Barney, J. B., and W. G. Ouchi. 1986. *Organizational Economics. Towards a New Paradigm for Understanding and Studying Organizations*. San Francisco: Jossey-Bass.

Bergström, E. 1980. *Projektorienterad marknadsföring*. Malmö: Liber.

Bingham, F. G., Jr., and B. T. Raffield III. 1990. *Business to Business Marketing Management*. Boston: Irwin.

Blau, P. 1964. *Exchange and Power in Social Life*. New York: Wiley.

Blau, P. 1968. "The Hierarchy of Authority in Organizations." *American Journal of Sociology* 73, pp. 453-467.

Blau, P. 1987. "Microprocess and Macrostructure." In *Social Exchange Theory*. edited by K. S. Cook. Beverly Hills: Sage, pp. 83-100.

Boddewyn, J. J. 1981. "Comparative Marketing: The First 25 Years." *Journal of International Business Studies*, Vol XII, (Spring/Summer).

Bond, M. H., ed. 1986. *The Psychology of the Chinese People*. Hong Kong: Oxford University Press.

Brunsson, N. 1982. "Företagsekonomi–Avbildning eller Språkbildning." In *Företagsekonomi–sanning eller moral*, edited by N. Brunsson. Lund: Studentlitteratur, pp. 100-112.

Bulmer, M. 1979. "Concepts in the Analysis of Qualitative Data." *Sociological Review*, Vol. 27, No. 4, pp. 651-677.

Carman, J. M., 1980. "Paradigms for Marketing Theory." In *Research in Marketing*, edited by J. Sheth. Greenwich, CT: Jai Press, pp. 1-36.

Chenery, H. B. 1979. *Structural Change and Development Policy.* Washington: World Bank.

Cheung, S. N. S. 1983. "The Contractual Nature of the Firm." *Journal of Law and Economics*, vol. 26, pp. 1-21.

Chisnall, P. M. 1989. *Strategic Industrial Marketing.* London: Prentice Hall.

Coase, R. H. 1937. "The Nature of the Firm." *Economica*, vol. 4, pp. 386-405.

Cohen, J. J., and M. C. Cyert. 1965. *Theory of the Firm: Resource Allocation in a Market Economy.* Englewood Cliffs, NJ: Prentice-Hall.

Cook, K. S., and R. M. Emerson. 1984. "Exchange Networks and the Analysis of Complex Organizations." *Research in the Sociology of Organizations*, vol. 3, pp. 1-30.

Corey, E. R. 1976. *Industrial Marketing: Cases and Concepts.* Englewood Cliffs, NJ: Prentice Hall.

Cundiff, E., and M. T. Hilger. 1988. *Marketing in the International Environment.* Englewood Cliffs, NJ: Prentice Hall.

DiMaggio, P., and W. W. Powell. 1983. "The Iron Cage Revisited: Institutional Isomorphism and Collective Rationality in Organizational Fields." *American Sociological Review* 48, pp. 147-160.

Eitel, E. J. 1984. *Feng Shui.* Singapore: Graham Brash.

El-Ansary, A. I. 1983. "The General Theory of Marketing: Revisited." Reprinted in *Marketing Theory. The Philosophy of Marketing Science*, edited by S. D. Hunt. Homewood, IL: Irwin, pp. 271-276.

Elster, J. 1986. *Vetenskapliga förklaringar.* Göteborg: Korpen.

Emerson, R. M. 1962. "Power-Dependence Relations." *American Sociological Review*, 27, pp. 31-40.

Ferrel, O. C., and J. R. Perrachione. 1983. "An Inquiry into Bagozzi's Formal Theory of Marketing Exchanges." Reprinted in *Marketing Theory. The Philosophy of Marketing Science*, edited by S. D. Hunt. Homewood, IL: Irwin, pp. 336-341.

Fligstein, N. 1985. "The Spread of the Multidivisional Form Among Large Firms, 1919-1979." *American Sociological Review*, vol. 50, pp. 377-391.

Gadde, L. E. 1980. *Distributionsstrategier på producentvarumarknader*, MTC no. 13, Stockholm: Liber.

Ghauri, P. 1983. "Negotiating International Package Deals. Swedish Firms and Developing Countries." Acta Upsaliensis, Studia Oeconomiae Negotiorum 17, Uppsala (dissertation).

Glaser, B. G. 1978. *Theoretical Sensitivity.* Mill Valley, CA: Sociology Press.

Glaser, B. G., and A. L. Strauss. 1967. *The Discovery of Grounded Theory. Strategies for Qualitative Research.* Chicago, IL: Aldine Publishing Co.

Haas, R. W. 1989. *Industrial Marketing Management* (4th ed.). Boston: PWS Kent.

Hadjikhani, A. 1984. "Organization of Manpower Training in International Package deal Projects." Department of Business Administration, University of Uppsala (dissertation).

Hägg, I., and J. Johanson. 1982. *Företag i nätverk-en ny syn på konkurrenskraft.* Stockholm: SNS.

Håkansson, H., ed. 1982. *International Marketing and Purchasing of Industrial Goods–An Interaction Approach.* Chichester: Wiley.

Håkansson, H., and C. Östberg. 1975. "Industrial Marketing–An Organizational Problem?" *Industrial Marketing Management*, vol. 4, no. 1/2, pp. 113-123.

Håkansson, H., and B. Wootz. 1975. *Företags inköpsbeteende.* Lund: Studentlitteratur.

Hallén, L. 1980. "Stability and Change in Supplier Relationships." In *Some Aspects of Control in International Business*, edited by L. Engwall and J. Johanson. Acta Upsaliensis, Studia Oeconomiae Negotiorum 12, Uppsala, pp. 83-101.

Hallén, L. 1982. "International Industrial Purchasing." Acta Upsaliensis, Studia Oeconomiae Negotiorum 13, Uppsala (dissertation).

Hallén, L., and J. Johanson, eds. 1989. *Advances in International Marketing*, vol. 3, Greenwich, CT: Jai Press.

Hammarkvist, K. O., H. Håkansson, and L. G. Mattsson. 1982. *Marknadsföring för konkurrenskraft.* Stockholm: Liber.

Hennart, J. F. 1982. *A Theory of Multinational Enterprise.* Ann Arbor: The University of Michigan Press.

Hill, R. M., R. S. Alexander, and J. Cross. 1975. *Industrial Marketing.* Homewood, IL: Irwin.

Hodgson, G. M. 1988. *Economics and Institutions. A Manifesto for a Modern Institutional Economics.* Cambridge: Polity Press.

Hofstede, G. 1980. *Culture's Consequences: International Differences in Work-Related Values.* Beverly Hills: Sage.

Hofstede, G. 1984. "Cultural Dimensions in Management and Planning." *Asia Pacific Journal of Management*, January, pp. 81-99.

Hofstede, G., and M. Bond. 1987. "Confucius and Economic Growth. New Insights into Culture's Consequences." Manuscript, Institute for Research on Intercultural Cooperation.

Howard, J. A. 1957/1963. *Marketing Management. Analysis and Planning.* Homewood, IL: Irwin.

Hunt, S. D. 1983. *Marketing Theory. The Philosophy of Marketing Science.* Homewood, IL: Irwin.

Jackson, B. 1985. *Winning and Keeping Industrial Customers.* Boston: D. C. Heath and Company.

Jansson, H. 1982. "Interfirm Linkages in a Developing Economy. The Case of Swedish Firms in India." Acta Universitatis Upsaliensies, Studia Oeconomiae Negotiorum 14, Uppsala (dissertation).

Jansson, H. 1984a. "Industrial Marketing Strategies of Transnational Corporations in Third World Markets." In *International Marketing Management*, edited by E. Kaynak, New York: Praeger, pp. 75-90.

Jansson, H. 1984b. "Vertical Firm Structures and Transnational Corporations in Developing Countries." In *Between Market and Hierarchy*, edited by I. Hägg, and F. Wiedersheim-Paul, Department of Business Administration, University of Uppsala, pp. 65-88.

Jansson, H. 1986. "Purchasing Strategies of Transnational Corporations in Import Substitution Countries." In *Advances in International Marketing*, vol. 1, edited by T. Cavusgil, Greenwich, CT: Jai Press, pp. 255-279.

Jansson, H. 1987. *Affärskulturer och relationer. En studie av svenska industriföretag i Sydöstasien*. MTC skriftserie nr. 29, Stockholm: Liber.

Jansson, H. 1988. *Strategier och organisation på avlägsna marknader*. Lund: Studentlitteratur.

Jansson, H. 1989. "Internationalization Processes in South-East Asia: An Extension or Another Process?" In *Global Business. Asia Pacific Dimensions*, edited by E. Kaynak and K.-H. Lee, London: Routledge, pp. 78-102.

Jansson, H. 1990. "Marketing to Projects in South East Asia." In *Advances in International Marketing*, vol. 3, edited by L. Hallén and J. Johanson, Greenwich, CT: Jai Press, pp. 259-276.

Jansson, H. 1992. "The Third Dimension of Global Organization. Transnational Industrial Corporations in Southeast Asia." Book manuscript, Institute of Economic Research, School of Economics and Management, Lund University.

Johanson, J., and L. G. Mattsson. 1987a. "Interorganizational Relations in Industrial Systems–A Network Approach." In *Strategies in Global Competition*, edited by N. Hood and J. E. Vahlne, New York: Croom Helm, pp. 287-314.

Johanson, J., and L. G. Mattsson. 1987b. "Interorganizational Relations in Industrial Systems: A Network Approach Compared with the Transaction-cost Approach." *International Studies of Management and Organization*, vol. XVII, no. 1, pp. 34-48.

Joy, A.J., and C. A. Ross. 1989. "Marketing and Development in Third World Contexts: An Evaluation and Future Directions." *Journal of Macromarketing*, Fall, pp. 17-31.

Kaynak, E. 1984. "Current Status of International Marketing Management." In *International Marketing Management*, edited by E. Kaynak. New York: Praeger, pp. 3-24.

Kogut, B. 1985. "A Critique of Transaction Cost Economics as a Theory of Organizational Behaviour." *Working Paper 85-05*, The Wharton School, University of Pennsylvania, Philadelphia, Pa.

Kotler, P. 1967/1984. *Marketing Management. Analysis, Planning and Control.* Englewood Cliffs, NJ: Prentice Hall.

Larsson, A. 1985. "Structure and Change. Power in the Transnational Enterprise." Acta Universitatis Upsaliensis, Studia Oeconomiae Negotiorum 23, Uppsala (dissertation).

Larsson, B. 1989. *Koncernföretaget*. Stockholm: EFI, Stockholm School of Economics.

Lasserre, P. 1988. "Corporate Strategic Management and the Overseas Chinese Groups." *Asia Pacific Journal of Management*, vol. 5, no. 2 (Jan.), pp. 115-131.

Leblebici, H. 1985. "Transactions and Organizational Forms: A Re-Analysis." *Journal of Organization Studies*, 6/2, s. 97-115.

Lecraw, D. J. 1984. "Pricing Strategies of Transnational Corporations." *Asia Pacific Journal of Management*, vol. 1, no. 2, pp. 112-119.
Lichtenthal, J. D., and L. L. Beik. 1984. "A History of the Definition of Marketing." *Research in Marketing*, vol. 7, pp. 133-163.
Lim, L. Y. C., and P. Gosling, eds. 1983. *The Chinese in Southeast Asia*. Singapore: Maruzen Asia.
Limlingan, V.S. 1986. "The Overseas Chinese in ASEAN: Business Strategies and Management Practices." DBA dissertation, Harvard University.
Lindberg, J. 1982. "Systemförsäljning." In *Exporthandboken*, edited by S. Söderman. Stockholm: Swedish Export Council.
Lip, E. 1989. *Feng Shui for Business*. Singapore: Times.
Lundgren, S., and G. Hedlund. 1983. *Svenska företag i Sydöstasien*. Stockholm: Institute of International Business Research, Stockholm School of Economics.
Mattsson, L. G. 1979. "Interorganizational Structures on Industrial Markets–A Challenge to Marketing Theory and Practice." Lecture at a Joint Berkeley Stanford Seminar in Marketing at Berkeley, Department of Business Administration, University of Uppsala (mimeo).
McCarthy, J. 1960/1975. *Basic Marketing*. Homewood, IL: Irwin.
McFarland, D. 1985. *Animal Behaviour*. London: Pitman.
McInnes, W. 1964. "A Conceptual Approach to Marketing." In *Theory in Marketing*, edited by R. Cox, W. Alderson and S. J. Shapiro. Homewood, IL: Irwin, pp. 51-67.
McManus, J. 1975. "The Cost of Alternative Economic Organizations." *Canadian Journal of Economics*, vol. 75, pp. 334-350.
Meyer, J. W., and W. R. Scott. 1983. *Organizational Environments. Ritual and Rationality*, Beverly Hills: Sage.
Mintzberg, H. 1987. "Crafting Strategy." *Harvard Business Review*, no. 4, pp. 66-75.
Mintzberg, H., and J. A. Waters. 1985. "Of Strategies, Deliberate and Emergent." *Strategic Management Journal*, vol. 6, pp. 257-272.
Myrdal, G. 1968a. *Asian Drama–An Inquiry into the Poverty of Nations*, vol. I-III, New York: Twentieth Century Fund.
Myrdal, G. 1968b. *Objetivitetsproblemet i samhällsforskningen*. Stockholm: Rabén & Sjögren.
Nicholas, S. 1986. "The Theory of Multinational Enterprise as a Transactional Mode." In *Multinationals–Theory and History*, edited by P. Hertner and G. Jones. London: Aldershot.
Palmer, D., R. Friedland, P. D. Jennings, and M. E. Powers. 1987. "The Economics and Politics of Structure: The Multidivisional Form and the Large U.S. Corporation." *Administrative Science Quarterly* 32, pp. 25-48.
Perrow, C. 1986. *Complex Organizations: A Critical Essay*, 3rd. ed. New York: Random House.
Pfeffer, J. 1981. *Power in Organizations*. Marshfield, MA: Pitman.
Porter, M. 1980. *Competitive Strategy. Techniques for Analyzing Industries and Competitors*. New York: The Free Press.

Porter, M. 1981. "The Contributions of Industrial Organization to Strategic Management." *Academy of Management Review* 6, pp. 609-620.

Porter, M. 1986. "Competition in Global Industries: A Conceptual Framework." In *Competition in Global Industries*, edited by M. Porter. Boston, MA: Harvard Business School Press, pp. 15-60.

Pye, L. W. 1985. *Asian Power and Politics. The Cultural Dimensions of Authority.* Cambridge: Belknap Press.

Redding, S. G. 1980. "Cognition as an Aspect of Culture and its Relation to Management Processes: An Exploratory View of the Chinese Case." *Journal of Management Studies*, 17, no. 2, pp. 127-148.

Redding, S. G., 1982. "Cultural Effects on the Marketing Process in Southeast Asia." *Journal of the Market Research Society* 24, no. 2, pp. 98-114.

Redding, S. G., and M. Ng. 1983. "The Role of 'Face' in the Organizational Perceptions of Chinese Managers." *International Studies of Management and Organization*, no. 3, pp. 92-123.

Redding, S. G., and D. Pugh. 1985. "The Formal and the Informal: Japanese and Chinese Organization Structures." Paper presented at The Enterprise and Management in East Asia Conference, Hong Kong.

Redding, S. G., and G. Y. Y. Wong. 1986. "Chinese Organizational Behaviour." In *The Psychology of the Chinese People*, edited by M. H. B. Bond. Hong Kong: Oxford University Press, pp. 267-295.

Reeder, R. R., E. G. Brierty, and B. H. Reeder. 1987. *Industrial Marketing. Analysis, Planning, and Control.* Englewood Cliffs, NJ: Prentice Hall.

Reve, T., and L. W. Stern. 1985. "The Political Economy Framework of Interorganizational Relations Revisited." In *Research in Marketing. Changing the Course of Marketing. Alternative Paradigms for Widening Marketing Theory*, suppl. 2, edited by N. Dholakia and J. Arndt. Greenwich, CT: Jai Press.

Robins, J. A. 1987. "Organizational Economics: Notes on the Use of Transaction-Cost Theory in the Study of Organizations." *Administrative Science Quarterly* 32, pp. 68-86.

Scott, W. R. 1987. "The Adolescence of Institutional Theory." *Administrative Science Quarterly* 32, pp. 493-511.

Smith, C., R. Whipp, and H. Willmott. 1988. "Case-Study Research in Accounting: Methodological Breakthrough or Ideological Weapon?" *Advances in Public Interest Accounting*, vol. 2, pp. 95-120.

Spradley, J. P. 1979. *The Ethnographic Interview.* Holt, Rinehart and Winston.

Stening, B. W., and J. E. Everett. 1984. "Japanese Managers in Southeast Asia: Amiable Superstars or Arrogant Upstarts?" *Asia Pacific Journal of Management*, vol. 1, no. 3, pp. 171-180.

Stern, L. W., and T. Reve. 1980. "Distribution Channels as Political Economies. A Framework for Comparative Analysis." *Journal of Marketing*, vol. 44, pp. 52-64.

Villacorta, W. V. 1976. "The Chinese in Southeast Asia: An Introduction." *Philippine Sociological Review*, vol. 24, no. 1-4, pp. 5-16.

Warren-Boulton, F. R. 1978. *Vertical Control of Markets. Business and Labor Practices*. Cambridge, MA: Ballinger.
Webster, F. 1979. *Industrial Marketing Strategy*. New York: Wiley and Sons.
Williamson, O. E. 1975. *Markets and Hierarchies: Analysis and Antitrust Implications*. New York: The Free Press.
Williamson, O. E. 1979. "Transaction-Cost Economics: The Governance of Contractual Relations." *Journal of Law and Economics* 22 (October), pp. 3-61.
Williamson, O. E. 1981. "The Modern Corporation: Origins, Evolution, Attributes." *Journal of Economic Literature* 19 (December), pp. 1537-68.
Williamson, O. E. 1985. *The Economic Institutions of Capitalism. Firms, Markets, Relational Contracting*. New York: The Free Press.
Williamson, O. E. 1986. "Economics and Sociology: Promoting a Dialog." *Working Paper 49*, Yale Law School.
Williamson, O. E. 1989. "Internal Economic Organization." In *Perspectives on the Economics of Organization*, edited by A. Malm. Lund: Institute of Economic Research, School of Economics and Management, Lund University.
Williamson, O. E., and W. G. Ouchi. 1981. "The Markets and Hierarchies and Visible Hand Perspectives." In *Perspectives on Organization Design and Behaviour*, edited by A. H. Van de Ven and W. F. Joyce. New York: Wiley, pp. 347-370.
Wimalasiri, J. 1988. "Cultural Influence on Aspects of Management: The Experience of the Chinese in Singapore." *Asia Pacific Journal of Management*, vol. 5, no. 3 (May), pp. 180-196.
Winship, C., and S. Rosen. 1988. "Introduction: Sociological and Economic Approaches to the Analysis of Social Structure." *American Journal of Sociology*, vol. 94, supplement, pp. 1-16.
Yin, R. Y. 1984/1989. *Case Study Research: Design and Methods*. Beverly Hills: Sage.

Index